Intracellular Staining of Mammalian Neurones

Biological Techniques Series

J. E. TREHERNE
Department of Zoology
University of Cambridge
England

P. H. RUBERY
Department of Biochemistry
University of Cambridge
England

Ion-sensitive Intracellular Microelectrodes, *R. C. Thomas*, 1978

Time-lapse Cinemicroscopy, *P. N. Riddle*, 1979

Immunochemical Methods in the Biological Sciences: Enzymes and Proteins, *R. J. Mayer* and *J. H. Walker*, 1980

Microclimate Measurement for Ecologists, *D. M. Unwin*, 1980

Whole-body Autoradiography, *C. G. Curtis, S. A. M. Cross, R. J. McCulloch* and *G. M. Powell*, 1981

Microelectrode Methods for Intracellular Recording and Ionophoresis, *R. D. Purves*, 1981

Red Cell Membranes—A Methodological Approach, *J. C. Ellory* and *J. D. Young*, 1982

Techniques of Flavonoid Identification, *K. R. Markham*, 1982

Techniques of Calcium Research, *M. V. Thomas*, 1982

Isolation of Membranes and Organelles from Plant Cells, *J. L. Hall* and *A. L. Moore*, 1983

Intracellular Staining of Mammalian Neurones, *A. G. Brown* and *R. E. W. Fyffe*, 1984

Intracellular Staining of Mammalian Neurones

A. G. Brown and R. E. W. Fyffe

Department of Veterinary Physiology
Summerhall
University of Edinburgh
Edinburgh, UK

1984

ACADEMIC PRESS

(Harcourt Brace Jovanovich, Publishers)

London   Orlando   San Diego   San Francisco   New York
Toronto   Montreal   Sydney   Tokyo   São Paulo

ACADEMIC PRESS INC. (LONDON) LTD
24—28 Oval Road,
London NW1

United States Edition published by
ACADEMIC PRESS INC.
(Harcourt Brace Jovanovich, Inc.)
Orlando, Florida 32887

British Library Cataloguing in Publication Data

Brown, A. G. (Alan Geoffrey)
    Intracellular staining of mammalian neurones
    1. Stains and staining (Microscopy)
    I. Title   II. Fyffe, R. E. W.
    578'.64   QH237

    ISBN 0-12-137220-0
    LCCCN 83-73143

Typeset by Bath Typesetting Ltd., Bath
Printed in Great Britain by Page Bros (Norwich) Ltd

Preface

Since the publication of our original report on intracellular staining of individual mammalian neurones with horseradish peroxidase (Snow *et al.*, 1976), we have been inundated with requests for demonstrations and with questions on particular aspects and techniques, such as: how to make microelectrodes for intra-axonal penetration, how to fill microelectrodes, what electrodes are "best" for certain applications, how to carry out the perfusion of the preparation and the histochemical reactions, etc. This book is in response to these requests and questions.

One of us is in print expressing reservations about publishing technical details (Brown, 1981). "Intracellular Staining of Mammalian Neurones" represents a retreat from that position and will, we hope, provide full details of the methods, so that the reader will be able to perform all the necessary steps to achieve success with the technique. The main reason for this retreat is the persisting demand by neuroscientists for details of our methods.

We should like to stress, however, that techniques change with time. We have indicated some of the changes that have taken place since 1975 and have described the methods as currently used in our laboratory. Doubtless, by the time this book reaches the reader there will have been some further changes, perhaps subtle ones or even more drastic ones; neuroscience is moving at a very rapid rate at the moment. But we do believe that this technical manual will provide a sound basis in the methodology of horseradish peroxidase staining of mammalian neurones for a considerable time.

We should like to thank our colleagues Drs P. K. Rose and P. J. Snow, with whom the method was originally developed, R. B. Hume who has modified the histological and histochemical processing over the years and prepared the figures for this book, and Zena McCubbin for typing the manuscript.

A. G. Brown

R. E. W. Fyffe

December, 1983.

Contents

1
Introduction to Intracellular Staining

I. INTRODUCTION

Two of the most important technical advances that have aided the experimental study of nervous systems have been the Golgi methods for staining nerve cells, allowing their visualization, and the introduction of micropipette electrodes (henceforth called microelectrodes, or simply electrodes, in this book), for intracellular electrophysiological recording from single neurones. The Golgi method (Golgi, 1873) and its subsequent modifications (including its development for electron microscopy) allow neuronal structure to be examined, particularly the organization and orientation of dendrites and, for some neurones, the trajectory and branching patterns of axons. Intracellular microelectrodes (Ling and Gerard, 1949) allow electrophysiological recordings to be made from within neurones, both from their soma and proximal dendrites and their axons, so that their functional properties may be studied. But, as powerful as these techniques are, they both suffer from certain disadvantages.

The Golgi methods are notoriously difficult and capricious. They stain neonatal neurones more efficiently than adult ones, and those neurones that do take up the stain do so in an indiscriminate manner, so that the experimenter may have to wait a long time before any particular type of neurone is stained. Some neuronal types may not take up the stain at all and, although traditionally it is believed that a successful Golgi impregnation will reveal all of a neurone, there is now serious doubt as to whether this is in fact the case (Somogyi and Smith, 1979). Except in special circumstances, the method is incapable of providing an identity for the neurones it stains. Thus, although α-motoneurones in the spinal cord may be identified by their position, size and axonal projection into the ventral roots, many other neurones in the cord remain anonymous except in terms of their location, size and dendritic organization. In order for an identity to be given to a neurone revealed by the Golgi method, it has to be favourably located or other evidence from various anatomical techniques (such as degeneration methods, ortho- and retrograde transport studies, histochemistry, etc.), as well as physiological information, has to be supplied.

A major drawback with intracellular microelectrode recording also concerns the identification of neurones. In the mammalian central nervous system, neurones are invisible to the experimenter (except in certain favourable circumstances, such as in the isolated retina and tissue slices) and identification depends on various indirect approaches. For example, a neurone with a long axon may be identified by antidromic activation. Neurones with short axons have to be recognized by more indirect means, such as the organization of their inputs (both excitatory and inhibitory). This sort of evidence often entails the accumulation of data over many years and is open to re-evaluation at any time. An example of this process is the identification of those spinal cord neurones responsible for Ia ("direct") inhibition of motoneurones (see Hultborn et al., 1971; Jankowska and Lindström, 1972).

Obviously, a technique that combines the power of the Golgi method with that of the intracellular microelectrode would be extremely useful, allowing both anatomical and physiological studies on single neurones, with the possibility of firm identification and light and electron microscopical analysis of neurones whose physiology was known. Such a technique is now available and it is the purpose of this book to provide the neuroscientist with details of its use. The method, the use of microelectrodes filled with horseradish peroxidase (HRP) for intracellular electrophysiology and subsequent staining of individual mammalian neurones, has opened up a new era in neuroscience. Like most techniques, it was developed out of a continuing search for new methods. The history of this development is briefly reviewed below; a fuller account of the history up to the early 1970s and the introduction of the Procion dye techniques can be found in the excellent book edited by Kater and Nicholson (1973).

II. HISTORY

An early problem faced by workers recording from single neurones in the mammalian central nervous system was that they did not know, except within very broad limits, from where their recordings were being taken. This was a problem found with both extracellular and intracellular recording techniques. Electrolytic lesions could be made by passing a relatively large current from a metal electrode, but they provided only crude localization. More effective, but still imprecise, was the deposition of iron from a steel electrode and its subsequent visualization in fixed tissue by the Prussian Blue reaction (Hess, 1932; Adrian and Moruzzi, 1939; Green, 1958).

Similarly, the Prussian Blue reaction for intracellular staining was reported by Kerkut and Walker (1962) but, although indicating which neurones had been recorded, they gave no additional information about neuronal anatomy. Subsequently, a number of dyes that could be ionophoresed from micro-electrodes were utilized, the most successful being Pontamine Sky Blue for extracellular marking (Hellon, 1971) and Fast Green for intracellular marking (Thomas and Wilson, 1966).

There was a significant advance in technique in 1968 with the publication of the Procion dye method (Kravitz *et al.*, 1968; Stretton and Kravitz, 1968). A number of these dyes were used, the most effective being Procion Yellow M4RS or M4RAN. When ionophoresed (or pressure injected; Remler *et al.*, 1968) into neurones, it can be visualized under the fluorescence microscope and reveals considerable neuronal anatomy (see below). More recently, another fluorescent dye, Lucifer Yellow CH, has been introduced (Stewart, 1978). This has certain advantages over Procion Yellow, especially for invertebrate neurones and mammalian central nervous system slices (see below). In 1972, Pitman *et al.* introduced the intracellular ionophoresis of cobaltous chloride. This method has had considerable success in invertebrate preparations, but has not proved successful in the mammalian central nervous system (see the book by Pitman in this series for full details of the cobalt techniques; Pitman, 1984).

An interesting intracellular staining method was introduced in 1968 by Globus *et al.* (see also Lux *et al.*, 1970a,b). This is the ionophoretic injection of tritiated glycine, with subsequent autoradiographic examination of the fixed and sectioned material. Considerable detail of the cell's anatomy can be visualized by this method, but it has not been taken up for general use. Undoubtedly, this is partly due to the availability of easier, more direct methods and the time required for producing the autoradiographs (5–6 weeks).

A watershed in the history of intracellular staining was reached in 1976, with the publication of a number of papers from different laboratories describing the intracellular injection of the enzyme horseradish peroxidase (HRP) (Cullheim and Kellerth, 1976; Jankowska *et al.*, 1976; Kitai *et al.*, 1976; Light and Durkovic, 1976; Snow *et al.*, 1976). HRP had been used previously as an anatomical tracer to delineate neuronal pathways, since its introduction for retrograde pathway tracing by LaVail and LaVail (1972, 1974). Also Graybiel and Devor (1974) used ionophoresis of HRP as an extracellular marker. Intracellular HRP ionophoresis produced striking Golgi-like light microscope preparations that showed the neurones' axons and collaterals in addition to detailed dendritic anatomy; furthermore, the HRP reaction product was electron dense and provided material for the electron micro-scopical analysis of recorded neurones. It is the purpose of this book to describe the techniques of intracellular staining with HRP, particularly as developed in our laboratory from one of the original methods (Snow *et al.*, 1976). In the remainder of this chapter, brief descriptions of the Procion and Lucifer dye methods, and the HRP method will be given and their major advantages and disadvantages discussed.

III. PROCION YELLOW

Procion Yellow is the best of the procion dyes for intracellular staining. The technique, introduced by Kravitz *et al.* (1968) and Stretton and Kravitz (1968), was modified for use in the mammalian central nervous system by Jankowska and Lindström (1970, 1971, 1972), by Barrett and Graubard (1970), Barrett and Crill (1974), and by Van Essen and Kelly (1973), and Kelly and Van Essen (1974). Microelectrodes filled with a 4–5% solution of Procion Yellow M4RS or M4RAN in distilled water are used. These electrodes, chosen appropriately for the job in hand, make good recording electrodes for responses and biophysical measurement. Intracellular ionophoresis is achieved by either constant hyperpolarizing currents or pulses of hyperpolarizing currents at up to 25 nA. The size of current that the electrode will pass depends on its resistance and the geometry of the tip. In general, a total of about 100 nA.min (it will be the standard practice in this book to express the amount of current passed as nA × min, i.e. nA.min) will provide satisfactory staining of large mammalian neurones, and small neurones will be well stained with considerably less.

After allowing a few hours (1–12) for the dye to diffuse throughout the neurone, the preparation is perfused with cold Ringer's solution and the appropriate parts of the tissue are removed as blocks and stored overnight in Ringer's solution at 2–4°C. Fixation is then carried out in 10% neutral buffered formalin. Alternatively, the preparation can be perfused with 10% buffered formalin (Barrett and Graubard, 1970). The fixed tissue can then be embedded in either celloidin or paraffin wax before cutting sections (15–20 μm), or frozen sections can be cut off the fixed tissue. In our hands, paraffin embedding gives the best results. Celloidin must be removed (with ether–alcohol), as it fluoresces. After clearing, the sections are viewed with the fluorescence microscope using UV illumination and appropriate filters (see Stretton and Kravitz, 1973; Van Orden, 1973; Jankowska and Lindström, 1970).

Material with Procion Yellow-filled neurones may be prepared for the electron microscope (Purves and McMahon, 1972; Kellerth, 1973; Berthold *et al.*, 1979). Useful data may be obtained with this technique, especially with regard to the ultrastructure of the cell body and proximal dendrites and the contacts they receive.

A. Disadvantages of Procion Yellow

Undoubtedly the major disadvantage of Procion Yellow as an intracellular stain for mammalian neurones is that it does not usually give a complete picture of the cell's dendritic tree. This has become obvious with the subsequent use of HRP. Furthermore, from its inception, it was apparent that Procion Yellow did not provide adequate staining of a neurone's axon. At best, only a few mm of axon are revealed and any initial axon collaterals are not filled, or are only stained at their origin.

The fact that Procion Yellow needs to be examined under the fluorescence microscope is a further disadvantage in comparison with HRP, which can be viewed under ordinary transmitted illumination. Furthermore, the fluorescence is relatively weak and may be difficult to resolve against background fluorescence. It is also difficult to make measurements of dendritic diameter in Procion Yellow filled neurones, as there is often a halo around the stain, especially when the dendrites are intensely stained. Repeated exposure to u.v. light causes fading of the fluorescence.

Although it is possible to recognize Procion filled profiles under the electron microscope, the identification of these profiles is much less easy than of those filled with HRP reaction product. Finally, Procion Yellow damages mammalian neurones (in contrast to invertebrate neurones which are apparently undamaged). Various types of damage have been reported by most workers (Barrett and Graubard, 1970; Jankowska and Lindström, 1970; Brown *et al.*, 1976). Again, this is in contrast to HRP, which does not usually give rise to untoward physiological effects.

B. Advantages of Procion Yellow

A major advantage of Procion Yellow is the relative ease with which the technique can be used. There is no histochemical processing as there is with HRP. If only the soma and major dendrites are of interest then Procion Yellow must remain an important choice and may be the prime choice for ultrastructural studies of these regions. Perhaps its most useful application is in combined studies, where dual staining of pairs of neurones is required. For example, one member of a pair might be stained with HRP and the other with Procion. This would allow the location of HRP-filled boutons on the soma and proximal dendritic tree of a Procion Yellow-filled neurone to be observed (see Snow *et al.*, 1976, and Fig. 39).

IV. LUCIFER YELLOW

In 1978, a more powerful fluorescent dye method for intracellular staining was introduced by Stewart. This was a naphthalimide dye called Lucifer Yellow CH. For invertebrate neurones Lucifer Yellow undoubtedly gives a more detailed picture than Procion Yellow, the fluorescence is more intense and finer processes are revealed, more of the axon and its branches may be seen and the dye spreads rapidly through the cell. So far, Lucifer Yellow has been used in the vertebrate for a variety of cell types in the retina and also in mammalian brain slice preparations (Cobbett and Cottrell, 1981; Knowles *et al.*, 1982). It does not appear to have been used successfully in *in vivo* mammalian preparations, except for dorsal root ganglion cells (Harper and Lawson, 1982).

The technique for intracellular staining with Lucifer Yellow is essentially the same as for Procion Yellow. Aqueous solutions of 3–5% Lucifer Yellow, or similar concentrations, in 1 M LiCl are used to fill microelectrodes and the dye is injected by hyperpolarizing current pulses of up to 20 nA. Tissue can be fixed with 4% phosphate buffered formaldehyde and embedded in glycol methacrylate treated with charcoal (Feder and O'Brien, 1968). Slides can be cleared and mounted in methyl salicylate (Cobbett and Cottrell, 1981). The tissue is viewed under u.v. light (see Stewart, 1978 for details).

A. Disadvantages of Lucifer Yellow

Except for the better quality of staining, Lucifer Yellow has similar disadvantages for the experimenter to Procion Yellow.

B. Advantages of Lucifer Yellow

As mentioned above, Lucifer Yellow provides a more complete picture of injected neurones than Procion Yellow. Otherwise the advantages are similar.

V. HORSERADISH PEROXIDASE

It is the purpose of this book to provide details of the HRP method. Here, a brief overview will be given and a summary of its disadvantages and advantages compared with the other methods.

Horseradish peroxidase (HRP) [donor: hydrogen-peroxide oxidoreductase; EC 1.11.1.7] is available from several commercial suppliers. We use Type VI from Sigma Chemical Co. The molecular weight is around 40 000 Daltons, and at least three isoenzymes are constituents. A haem group is essential for its enzymatic activity and it is active over a wide range of pH values.

HRP catalyses the reduction of appropriate hydrogen donors in the presence of hydroperoxides. It is very specific for the hydrogen acceptor: H_2O_2 is by far the best and is always used in histochemical procedures. It is far less specific for the hydrogen donor: a large variety of substances can be employed. Hence the variety of methods currently being used and developed in the quest for sensitive and specific histochemical procedures.

HRP cannot be satisfactorily ionophoresed from a simple aqueous solution. The microelectrode must contain some inorganic ions in addition to the HRP. The solution should be buffered to maximize electrophoretic mobility. Since its most active constituent, isoenzyme C (Bunt et al., 1976), has an isoelectric point around pH 9.0, buffering in the pH range 7.3–8.6 is commonly chosen. Several recipes are available: we use a 4–10% solution of HRP in 0.1 M Tris/HCl containing 0.2 M KCl (pH 7.6–8.6); Jankowska et al. (1976) used a 15–20% solution of HRP in 0.2 M NaOH (pH about 8.5). For further details of solutions see Chapter 3. HRP is injected into neurones by passing positive (depolarizing) current through the electrode, either constant DC current or pulses of various durations, in the range from a few to about 20 nA in amplitude. The amount of current needed depends on the size of the cell (see Chapters 4 and 5).

Between one and nine hours after injection, the animal is perfused with saline followed by fixative (formaldehyde or glutaraldehyde/formaldehyde mixtures, Chapter 6). Sections can be cut on either the freezing microtome or on a tissue slicer (Chapter 6) and then reacted for HRP by one of a number of methods: using diamino-benzidine, pyrocatechol/p-phenylenediamine or tetramethylbenzidine (Chapter 6). Sections can be viewed in the light microscope, with light or dark field condensers, or in the electron microscope.

A. Disadvantages of HRP

There are few disadvantages. Perhaps the main one is that if the enzyme leaks from the tip of the microelectrode then an extracellular blob of reaction product, whose size depends on the degree of spilling, results. This can be annoying and often leads to spoiling a certain amount of tissue (usually at the site of a cell you would particularly have loved to stain). Spillage of the fluorescent dyes causes no problems as they seem to be removed from, or diffuse away from, the extracellular space.

Because of the sensitivity of the HRP method, another possible disadvantage is that if neurones are stained in close proximity to one another, say within a few mm, it might be difficult to differentiate the processes of one from those of the other. One soon learns to avoid staining cells too close together and, of course, it may be necessary or advantageous to stain near neighbours in some experiments (e.g. Brown et al., 1980).

Although HRP has to be reacted histochemically for visualization, this reaction is remarkably robust and rarely causes problems (see Chapter 6).

B. Advantages of HRP

These are numerous. The most important advantage is that HRP provides the most complete view of a neurone that can be attained at present. In fact, it gives a more complete view than the Golgi method which is now known sometimes to miss complete dendritic systems (Somogyi and Smith, 1979; see also Brown and Fyffe, 1981). Up to 1 cm or so of a neurone's axon is stained together with many of its collaterals to their terminal boutons. In this latter respect, HRP is at least an order of magnitude better than the fluorescent dyes.

In addition to providing a Golgi-like picture of a neurone at the light microscope level, excellent electron micrographs of stained neurones can be obtained (Jankowska *et al.*, 1976; Cullheim and Kellerth, 1976), thus allowing the ultrastructure of electrophysiologically recorded neurones to be studied (see Chapter 6).

The sharp contrast provided by HRP reaction product makes quantitative measurements of dendritic lengths, diameters, axon diameters, bouton dimensions, etc. very straightforward. Furthermore, there is very little fading of the reaction product over several years if the sections are kept in a box (the pyrocatechol/*p*-phenylenediamine reaction seems particularly free from fading and we have material more than 5 years old that is as good as new).

2
General Techniques for Intracellular Recording and Current Passing

I. MAINTENANCE OF MAMMALIAN PREPARATIONS

It will doubtless seem obvious to most readers that before any satisfactory electrophysiological recordings can be made from mammalian neurones it is necessary to have a preparation in as good a physiological condition as possible. It is our contention, however, that insufficient attention to this problem is the cause of many failures and poor results. It is imperative that time and trouble be spent on obtaining consistent preparations in excellent condition.

Reduction of trauma during surgery is important, but quick and relatively rough surgery produces better results than painstaking surgery that takes a long time. Any unavoidable blood loss should be made up promptly, minor losses being replaced with normal saline as fluid loss is more important than loss of blood *per se* and larger losses should be made up with a plasma expander (e.g. dextran solutions, but beware of using too much with subsequent rouleaux formation) or, if available, compatible blood. Air embolism is a risk in preparations in which the skull is opened either for access to brain targets or for decerebrations etc. All exposed bone edges should be plugged with bone wax, dental cement or some suitable alternative, such as modelling compounds like Plasticene.

The preparation should, of course, be kept at normal body temperature with either infra-red lamps or, more conveniently, with an electric blanket, except where this would severely limit access. Any heating system should include servo control of deep body temperature. Large exposures of tissue, such as occur with a standard lumbosacral laminectomy, for example, lead to loss of heat. It is often useful to have separate heating control of large paraffin pools.

Various physiological parameters should always be monitored. These include arterial blood pressure, end-tidal CO_2, ECG (electrocardiogram) and EEG (electroencephalogram). We monitor the first two of these routinely.

Blood pressure measurements are important and it is convenient to have a continuous chart record to allow variation over time to be observed; a gradual decline from an acceptable value would indicate deterioration, a gradual rise in an anaesthetized and paralysed preparation might indicate that the anaesthetic level was lightening. Experiments are often terminated if systolic pressure falls below a certain value, such as 70 or 80 mmHg, and some workers use a value of 100 mmHg as their cut-off point and terminate if the systolic pressure falls below this value for more than a certain time. Diastolic pressure is also important, as it indicates the peripheral resistance: an animal may have a systolic pressure that is acceptable and yet the diastolic may be far too low. End-tidal CO_2 is a convenient measure of the general acid–base balance of the animal. It can be kept within normal limits by altering the rate and depth of respiration, if the preparation is being artificially respired, or by injections of bicarbonate solutions if end-tidal CO_2 is low. It could be argued that monitoring the blood directly for either P_{CO_2} or P_{O_2} and pH would provide better information about both the oxygenation of the blood and the acid–base balance. These measurements are rarely made except when the neurophysiology of respiration is being studied. Similar measurements of the cerebrospinal fluid should also give a better indication of the state of the extracellular fluid bathing the central nervous system.

It may be appropriate to monitor either ECG or EEG in some experiments. We do not make these recordings routinely. EEG may be useful for assessing anaesthetic level and the general condition of the cerebral cortex. In paralysed anaesthetized preparations the degree of constriction/dilatation of the pupils of the eyes can give information about the degree of stress of the preparation. If all other parameters are normal and the eyes are dilated a small additional dose of anaesthetic is often indicated.

Finally, we routinely catheterize the bladder and keep it empty. Apart from providing information on kidney function and the general state of hydration of the preparation, it also helps to stabilize the blood pressure and avoid blood pressure surges due to an overfull bladder. Continuous production of dilute urine should be aimed for and injections of saline (i.v.) made in order to maintain this state.

II. PROVIDING STABILITY OF CENTRAL NERVOUS SYSTEM (CNS) STRUCTURES

A. Basic Apparatus Requirements

An essential requirement for intraneuronal recording from mammalian preparations is that the target neurones should not be subjected to movement caused by external or internal influences. The laboratory should have a stable floor or the experimental table should be on anti-vibration mountings. The importance of this requirement cannot be over-emphasized. There is nothing more annoying than to impale a cell, often after many hours of searching, only to dislodge the electrode by knocking the experimental table. Even with a massive table properly situated it is still possible to dislodge intracellular electrodes and care to avoid unnecessary mechanical disturbance will still be required. The brain or spinal cord can usually be adequately stabilized with one of the commercially available head holders or spinal frames, although additional stability from internally generated movements (respiratory and cardiovascular) will also be required (see below).

B. Spinal Cord

The most effective way to stabilize the spinal cord for intracellular micro-electrode work is to stretch the spinal column between two fixation points that flank the region of interest. For the lumbosacral cord of cat the best points are: (i) the iliac crests, which are gripped by hip pins (it is not necessary to expose the bone, the pins can be fixed with the skin interposed); and (ii) the dorsal spines of the first two lumbar vertebrae or the last thoracic and first lumbar vertebrae taken together (by taking two spines together any flexion at the fixation point is prevented and more traction can be placed on the spinal column when it is stretched). For work on the thoracic or lower cervical cord, fixation on two sets of spines is effective, and for work on the upper cervical cord and the lower medulla, fixation can be by ear-bars and one set of dorsal spines. The animal frame we use is shown in Fig. 1.

Some further aids to stability are usually necessary. In our experience, the use of a bilateral pneumothorax is generally the only additional procedure necessary for cats. We insert rigid plastic tubes into the pleural cavity through the intercostal spaces. The external ends of the tubes are closed with thin-walled balloons (made from the fingers of disposable surgical gloves). Closing the tubes in this way prevents fluid loss by evaporation from the animal and the respiratory movements are transferred to the balloons. It is sometimes necessary to cut the ends off the balloons to remove the final degrees of respiratory pulsation. If the balloons are opened in this way then care must be taken that the animal does not become dehydrated over the long duration of the usual type of neurophysiological experiment.

Many laboratories use additional means for obtaining stability. As mentioned above, we do not find them necessary in the cat, but they may help other preparations, or indeed may be necessary if for some reason satisfactory stability cannot be achieved. (i) The formation of a dural hammock is a quite common practice: the dura is stitched to either surrounding muscle or the animal frame by loops of thread. In this way the cord is lifted

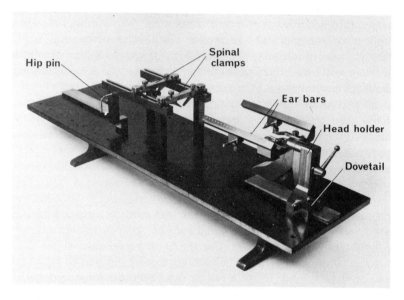

Fig. 1. Spinal frame and stereotaxic head holder. This is the frame used by the authors. The two sets of vertebral spine clamps and the pins for fixing the pelvis are mounted on the spinal frame. The head holder is free to move along the dove-tail. The apparatus was made by R. Clark in our department.

away from the underlying vertebral bodies and pulsating blood vessels. Care has to be taken that the lifting of the spinal cord does not lead to restriction of blood flow to the cord or obstruction of the venous drainage. A similar effect to a dural hammock can be achieved more easily by placing small pledgets of cotton wool (soaked in saline and then wrung out) between the dura and the bone or a platform may be used to support the cord or, for example, a dorsal root ganglion. The same care has to be taken not to restrict blood flow. (ii) Flooding the exposed spinal cord with a warm agar–saline mixture which then sets can be quite effective. A 4% solution of agar in normal saline is warmed to melt the agar and poured onto the cord at just above the setting point. For best results a wide cord exposure is preferred so that the agar forms a fairly large but shallow pool, only a mm or two deep over the highest parts of the cord. Such a shallow pool allows direct visual control and microelectrodes can be advanced through the agar without too much risk of blocking. However, it will usually be necessary to make holes in the pia-arachnoid for the microelectrode before pouring on the agar–saline. The agar–saline can be easily removed at any time but it is important not to let the exposed tissue dry out. If paraffin oil is allowed to come into contact with the tissues at any time then it will be impossible to use agar-saline thereafter: the oil forms a slippery surface between the agar and the cord and it is impossible to obtain fixation. Where the exposure is relatively restricted but deep, e.g. over the medulla or upper cervical cord, a deep agar "pool" may be necessary. Here it will be impossible to see through the thick layer of agar. To avoid frequent removal of the agar a useful trick is to place a modelled block of Plasticene (or other suitable material) on the area of interest, flood the area with agar-saline and, when the agar has set, remove the Plasticene. The hole in the agar can be filled with saline and the microelectrode advanced through it. (iii) Pressure feet can also be used to prevent movements. It is best to use clear plastic for these to allow visual inspection of the blood flow underneath as their greatest drawback is that they interfere with blood flow. They can be used in conjunction with an agar-saline method. A hole, or series of holes, in the plate allows access for the microelectrode.

C. Brain

When structures deep in the brain are being studied movement problems are rare. Any standard stereotaxic head holder will hold the skull satisfactorily. Pulsations due to cardiovascular or respiratory causes are only troublesome when recording from superficial structures (cerebral and cerebellar cortices). Agar–saline and/or pressure plates may be used, and great care will be needed to prevent damage to the superficial neurones. The best method of avoiding movement problems is to recreate the conditions pertaining before the skull (and dura) was opened, by the use of a closed chamber technique. A suitable chamber (that can be used in conjunction with a stepping-motor driven micromanipulator) is that described by Andersson and Källström (1971).

III. SPECIFIC APPARATUS REQUIREMENTS

A. Introduction

The essential link in both the intracellular recording of neurophysiological events and the intracellular injection of HRP is the microelectrode. This tool is so important that we have given a complete chapter to its manufacture and testing (Chapter 3). There are other important apparatus requirements for successful recording and HRP injection, however, such as a suitable electrometer for coupling the electrode to the main amplifiers and for applying current to the electrode, and a suitable current source for the pulses. These will be discussed here.

B. Microelectrode Holder and Micromanipulator

An electrode holder mounted on a micromanipulator will be required. Electrode holders are, nowadays, usually purchased as part of an electrometer system and incorporate its input stage. Micromanipulators can vary from simple hand-operated devices to more sophisticated stepping-motor or piezo-electric instruments. Our present manipulator is a stepping-motor type based on a design by Clark and Ramsay (1975) and allows vertical movement that can be either continuous, with varying rates, or in steps down to 2 μm. Microelectrode depth is displayed digitally. Providing controlled abrupt movements, such a stepping-motor drive allows precise tracking with the electrode and, most importantly, aids intracellular penetration. The

Fig. 2. Microelectrode carrier and stepping-motor drive. The microelectrode carrier is mounted on an arc which is, in turn, carried on a lathe movement. For further description of the arc and the lathe movement, see the text. The stepping motor is visible above the arc. The electrical leads are not connected to the motor. This apparatus was made by R. Clark in our department (see Clark and Ramsay, 1975).

yield of high quality intracellular penetrations increased enormously once we had installed the stepping-motor drive (in place of a previous hand-operated micromanipulator) and the drive is of the greatest importance to our success.

The stepping-motor drive and electrode holder assembly needs to be mounted so that movements in directions other than the vertical can be controlled. We use a system modified from that of Eide and Källström (1968) by Clark and Ramsay (1975). The micromanipulator is carried on an arc (Fig. 2) on which it can be moved to allow penetrations at an angle within the vertical plane. The arc itself can be rotated about an axis in its own plane so that angled approaches to the vertical plane may be made. Finally, the arc is carried on a lathe movement that permits displacements of the whole assembly parallel to the length and width of the experimental table and animal frame. The lathe movements can be controlled down to about 10–20 μm. Using this manipulator, approaches to any central nervous system target are possible, usually from more than one direction.

C. Electrometer

An electrometer with bridge balancing and current passing capabilities will make life easier. It is possible to use a system where the recording amplifier is switched off during current passage (for HRP injection), for example by switching to an external current source that by-passes the first stage amplifier. But this leads to the loss of important monitoring information. The ability to monitor the condition of the neurone during the passage of ejection current is, in our opinion, vital to consistent success with the HRP method. Not only does such a facility allow one to ascertain the condition of the neurone, it also allows assessment of the current passing capabilities of the microelectrode. General details of suitable amplifiers will be found in the book by Purves (1981) in this series. Here we give some information about their use for HRP injection.

1. Low voltage electrometer with bridge balance

This type of electrometer is ideal for use with microelectrodes, having relatively low impedances of the order of 20–60 MΩ. Because of the low voltage input, it is difficult to pass more than a few nanoamps of current if the electrode resistance is greater than about 60 MΩ. Increasing resistance (usually due to plugging or polarization of the tip) manifests itself in blocking of the electrometer. Thus, the amplifier becomes saturated, no biological signals (including the potential changes at the electrode tip) are recorded and ionophoresis usually ceases.

As mentioned above, these electrometers may incorporate "breakaway" circuits, whereby the amplifier input is disconnected and the current source is applied directly to the microelectrode. High currents can be attained, but there is serious loss of signal monitoring. We seldom use this mode, as it is much more convenient simply to change the blocked electrode to one which is more compatible to the input characteristics of the electrometer.

During current passage (pulses), it is advisable to make use of bridge balancing circuitry to attempt to balance the DC potential. First, this permits monitoring of all neural events, including those occurring during or even because of the current injection. Second, even if only the initial part of a (long) current-induced voltage change can be adequately balanced, the slope of the potential change gives an indication of the degree of polarization or plugging at the electrode tip. If it rises too quickly, saturation and failure of the ionophoresis will occur.

The commercially available electrometer model we use, with 20–80 MΩ electrodes, has a peak to peak noise level of 0.5–1.0 mV when no special care is taken to remove power supply or grounding induced noise sources. These can of course be reduced with a little more effort in laboratory design etc., but the noise level is satisfactory for both extracellular and intracellular recording (of action potentials and naturally or electrically evoked composite excitatory or inhibitory post-synaptic potentials). Events produced by single afferent axons, or occasionally some distant or very small extracellular potentials, require averaging techniques or a better signal to noise ratio (see Purves, 1981).

2. High voltage electrometers

Jochem et al. (1981) developed an electrometer which overcame some of the current passing limitations of the amplifier and allowed the use of fine-tipped (high resistance) microelectrodes. Full details of their circuitry are published in their paper, and more recently, the electrometer has been modified to give lower noise levels (see Metz et al., 1982). This type of electrometer was central to some of the outstanding achievements in the last few years, such as the staining of small substantia gelatinosa neurones and, more impressively, the staining of fine diameter myelinated afferent fibres (Light and Perl, 1979; Réthelyi et al., 1982). These applications, especially the latter, necessitated the use of fine-tipped microelectrodes and only with the high voltage electrometer could enough current be passed during the fairly short-lived intracellular or intra-axonal penetrations. These electrometers also have facilities for bridge balancing, current monitoring and digital display of potential changes or microelectrode resistance.

D. Current Source and Current Monitoring

An external source of current will be required. Again, the reader is referred to Purves (1981). For HRP injection, with currents of 2–20 nA, it is not so vital to have a sophisticated current pump as it would be for making biophysical measurements; a fairly simple voltage source will suffice. However, the ability to trigger the injecting current and to control its amplitude, polarity and shape are all advisable and when combined with an electrometer with bridge balancing and current passing capabilities the system will be easy to use and, more importantly, to control. A means of monitoring the amount of current being passed is essential.

In the published literature, reference is usually made to the current intensity (amplitude in nA) used and often to the duration of the injection. The current and time factors are combined to indicate, in a nominal sense, the "size" of the injection in nA mins—purely arbitrary units. These units are, however, quite useful when comparing a series of attempts at staining cells or axons of a given population. We, and others, find with HRP that a high current and short total duration is usually more effective than low current for a long time; depending of course on the size of the cell or axon being studied.

Current can be measured accurately by observing the voltage drop across a resistor in series with the microelectrode. Most of the commercially available electrometers do not use this method, and consequently some of the "monitored" currents do not give accurately the current passing through the microelectrode, but simply reflect the applied voltage. However, as long as the voltage is within the electrometer input limitations and the electrode is not blocked, these values are a useful indication of the ionophoretic current.

That "some" current is flowing is always indicated by membrane potential changes and other current-evoked responses (ON- or OFF-spikes) when the current passes some threshold level. The occurrence of ON- or OFF-responses is a useful indicator of the current polarity: depolarizing current (+ve) generates ON-spikes; hyperpolarizing current (−ve) will often produce OFF- (or rebound) responses.

3
Microelectrodes for Horseradish Peroxidase Staining

I. INTRODUCTION

The "magic" of microelectrode manufacture permeates almost all discussion between experimentally active neuroscientists. Everyone wants to know what "tricks" their colleagues use. This chapter will describe in detail our very straightforward techniques for pulling, filling and preparing microelectrodes for intracellular or intra-axonal ionophoresis of HRP. It is worth stating now that our electrodes are suitable for most applications in the spinal cord (except perhaps small myelinated nerve fibres—see Light and Perl, 1979). They also function satisfactorily in "higher" centres (as used by Martin and Whitteridge, 1982). Advice to beginners would be: (i) if intracellular studies (recording only) are already under way in the laboratory then continue with the same type of electrode and simply fill it with HRP; (ii) if a novel system is to be attacked, with no prior experience of intracellular work, then follow the directions in this book (we accept no responsibility for the results) and improvise.

II. MAKING MICROELECTRODES

A. The Microelectrode Puller

The electrode puller is regarded by the experimenter as perhaps the most vital and personal piece of equipment in the laboratory. Within the discipline of neuroscience it is used to produce the final tool for intracellular recording and staining, extracellular recording, ionophoretic application of drugs, biophysical measurement including patch clamp techniques, etc. These complex and various applications mean that no one type of microelectrode is suitable for all; however, experimental design and appropriate choice of electrode can permit combinations of, for example, intracellular staining with basic biophysical measurements.

Ideal microelectrode characteristics have been defined by some authors, in terms of length, taper, tip size, rigidity, resistive and capacitive properties, ease of filling with electrolyte, etc.; the ideal microelectrode puller would have adjustable controls to provide for consistent production of predetermined microelectrodes. The ideal puller does not exist, but in the last few years several new designs have been developed which do go some way to giving additional control over the electrode manufacture. Important among this new generation of pullers are the model described by Brown and Flaming (1977) and that described by Ensor (1979). Both of these are discussed elsewhere (Purves, 1981). It should be noted that the aim of both of these designs was to produce microelectrodes with very fine tip diameters and relatively short shanks. Theoretically, these are "good" electrodes. Indeed, such

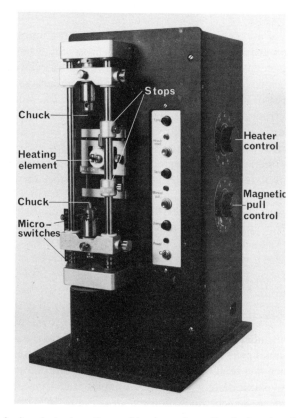

Fig. 3. Vertical microelectrode puller used by the authors. For further description see the text. The puller was made by R. Clark in our department.

electrodes are essential, or at least advantageous, for staining very small neuronal elements and have been used to great effect in some studies on spinal cord, retina and cerebral cortex. These will probably be the pullers and electrode types of choice in future developments.

Until now, we have used a much simpler design (Fig. 3), which can be built relatively cheaply. This is a vertical puller of the type originally described by Winsbury (1956) and detailed as his machine of use by Thomas (1978) in this series. It was designed by R. Clark for Forth Instruments. There are only one or two modifications in our model: the solenoid is switched on automatically by microswitches whose position can be adjusted to give longer or shorter gravity pulls of the softening glass; further, we use nichrome filament (Tophet A 1.185 Ω yd^{-1}) as the heater rather than steel or platinum coils. The nichrome is more rigid than platinum and can be formed into the required shape readily. It oxidizes fairly slowly so that one filament may last for two to three months of use. The deterioration means that adjustments must, however, be made regularly to the heater current amplitude.

The puller can be used to make multi-barrel assemblies; up to the present, we have only used single barrel microelectrodes. Whilst optimally shaped electrodes are produced by trial and error, the following guidelines will indicate the sort of parameters we pay attention to in their manufacture.

B. Choice of Glass and Electrode Manufacture

There is now a wide variety of commercially available glass for micro-electrode manufacture, ranging in external and internal diameter sizes, presence or absence of internal filaments, theta tubing, thin wall glass, etc. Our choice is 1.5 mm O.D. silicon glass containing an internal filament. The physics of glass pulling and the effect of the inner filament is discussed by Purves (1981). For our purposes the filament is an aid to filling the electrode with the HRP solution. We take no special precautions in storing or cleaning the glass before pulling. A suitable length is simply placed within the heating filament and clamped at top and bottom. The bottom clamp is free to slide down the parallel vertical runners of the puller. The heater switch is activated, the variable control having been previously adjusted. If possible, a draught shield should cover the puller at this stage. The glass begins to soften and the weight of the bottom clamp begins to pull it down. After 2–6 mm the solenoid is activated to provide the final pull to separate the glass. The process takes a few seconds.

In our experience, both electrodes formed from the one piece of glass are similar in their major characteristics and perform equally well in the experiment. Like the standard horizontal pullers, but not the Brown and Flaming or Ensor models, the vertical puller produces rather long electrodes. We have found no disadvantage in this. The shank length is of the order of 10–16 mm. The taper of the shank is gradual and there is usually a fairly abrupt change in taper to form the tip. So long as the shank is not too long (> 16 mm) or parallel sided we use it. All the electrodes made are visually examined for straightness and taper. Electrodes of 14–16 mm usually have more uniform taper than our shorter ones and we have tended to use these recently. We have never subjected any of our electodes to electronmicroscopic examination because, as we will discuss later, our final choice of electrode is based on how well it works, not on tip size or resistance measurement.

Generally, 12–20 electrodes are made for each experiment. They are pulled two to three days before use. Storage and filling are described in the following sections.

C. Filling the Electrode

We use Sigma type VI HRP. A 9% w/v solution is made from 9.0 mg of the solid enzyme dissolved in 0.1 ml of Tris HCl buffer. The buffer is 0.1 M Tris/HCl and contains 0.2 M KCl at pH 8.6 (Table 1). The pH of similar solutions used in other laboratories ranges from 7.3–8.6. The HRP may be dissolved in NaOH (Jankowska *et al.*, 1976). The Tris buffer solution is kept at 4°C before use. We do not filter the buffer or the final HRP solution (the volume is too small anyway) although care is taken to ensure that all of the solid HRP is dissolved.

Table 1. Tris-HCl buffer for dissolving HRP and filling microelectrodes (0.2 M HCl, 0.05 M Tris, 0.2 M KCl)

(i)	3.03 g Tris buffer 7.46 g KCl } dissolved in 450 ml distilled water
(ii)	Add 0.2 M HCl to give pH of 8.6
(iii)	Make up to 500 ml with distilled water

The electrodes are filled soon after manufacture, on the day prior to the experiment. The HRP solution is made up just before use. A fine polythene cannula is inserted down to the shoulder of the electrode and enough HRP solution is introduced to fill the tip, shank and about 5 mm of the body of the electrode. The cannula is removed and the electrode is placed vertically (tip down) in a container jar which has some distilled water in the base. The electrode tips do not touch the water. They are then stored, with a lid on the jar, in a refrigerator overnight. Immediately before use, the barrel of the electrode may be filled with buffer or other electrolyte. This is not necessary of course if connexion to the input stage of the eletrometer is via a silver wire which can be inserted down into the HRP solution in the tip of the electrode.

The problem of filling electrodes with HRP is that the solution, especially at the concentrations used by various workers so far (4–30%), is viscous. Even with fibre filling methods, air bubbles form in the shank and often occur near the tip. Many of the bubbles disappear during the overnight storage. The ones which remain can be easily removed either by manipulation with a fine whisker (cats' are best!) or by applying negative pressure above the electrode. We use a suction method: each electrode is linked via a connector (a Tuohy-Borst adaptor with female Luer lock, available from Becton Dickinson UK Ltd., York House, Empire Way, Wembley, Middlesex) to a 20 ml syringe and suction applied until bubbles reach the electrode shoulder. The syringe barrel is released, the size of the bubble decreases and it is able to migrate to the surface of the solution. It is somewhat time-consuming (half an hour for a batch of 15–20 electrodes) but worthwhile, especially if bubbles are present near the tip. Bubble removal can be carried out during other activities, such as electrode bevelling (see later). Placing the whole batch of electrodes in a vacuum jar and evacuating the jar is a quicker method.

For very fine microelectrodes, a useful method is to fill the tip and shank initially with a very small drop of the Tris buffer solution. After filling is complete (several hours to be on the safe side), and it is clear that no bubbles remain then the HRP solution ($> 10\%$ is best) is introduced as described above to the shoulder of the electrode. Two to three days are sufficient to effect filling of the tip with HRP by diffusion. Again, the electrodes are best stored vertically, with tip down, in a moist atmosphere and at about 4°C.

D. Preparation of the HRP-filled Microelectrode for Use

Our electrodes have AC resistances (1 KHz square wave) of 80–140 MΩ when tested in 0.9% saline. The fine tipped electrodes used by others range from 120–240 MΩ, in general, and may be higher. We have not felt it necessary to try to compare the parameters of our HRP-filled electrodes with similar ones containing standard electrolytes, such as 3M KCl. Ability to penetrate cells and pass current are the important considerations. Generally, however, the freshly filled electrodes will show about three times the resistance when filled with HRP/Tris buffer compared with KCl. Whilst some electrodes (e.g. fine tipped ones) may be used without further attention, we bevel our electrodes before use. This reduces their resistance (to make current passing easier) and creates a sharper tip (to facilitate cell or axon penetration), either by true bevelling or by producing freshly broken edges at the tip. The principles and methods of bevelling have been described in detail by Barrett and Whitlock (1973). Since then, other methods have developed. These include the grinding technique of Brown and Flaming (1975), the jet-stream method (Ogden *et al.*, 1978; Corson *et al.*, 1979), especially useful for fine electrodes, and the slurry beveller (Lederer *et al.*, 1979), which also suitably bevels fine electrodes. Our method is similar to Brown and Flaming's and uses a commercially available Narishige EG-5 microgrinder. The grinder surface is 0.05 μm AlO_3 particles embedded in a polyurethane base thinly spread on a clean glass disk. Several disks can be made at a time, and they can be stored for many months once hardened (after 24 h at least). Once in use, and treated carefully, they may last for 10–12 experiments before deteriorating. To monitor electrode resistance and the bevelling process, we include an electrode test circuit in our bevelling equipment. Contact and grounding is made by a cotton wick soaked in 0.9% saline and just touching the bevelling surface. Electrodes to be bevelled are mounted in a manipulator and advanced at 45° towards the rotating glass plate. The direction of approach follows the direction (anti-clockwise) of spin of the disk. Under microscopic control, the tip of the electrode is brought into contact with the thin layer of saline deposited on the surface by the wick. The moment of contact is apparent visually by observing the formation of a meniscus of fluid around the tip and, simultaneously, by the appearance of an oscilloscope trace, monitoring the electrode's resistance. This starts off at about 80–140 MΩ. The electrode is then advanced in 2 μm steps, until the resistance begins to drop, or high frequency noise is superimposed on the square wave electrode test trace. This occurs as the tip makes contact with particles of the grinding powder. We continue advancing the electrode slowly, the resistance will usually decrease in finite steps of 2–10 MΩ. When a final resistance of 15–60 MΩ is attained the electrode is removed and checked on the microscope. At this point, there are two main features to examine. (i) Has the electrode been bevelled too much or simply broken? A tip diameter of more than 2 μm makes the electrode unsuitable for most purposes. Related to this is the possible leakage of HRP from tips which are too large. (ii) Have bubbles or particulate matter been introduced into the tip during the bevelling process? If affirmative answers to these questions arise, then discard the electrode. The others proceed to the ultimate test—do they work and provide results in the preparation?

At this point, we would emphasize that the physical parameters of the electrode (length, taper, tip size, resistance) we have monitored so far are simply guidelines which experience tells us provide good electrodes. Each experimenter will find "good" electrodes peculiar to his or her requirements and will then *make* them work. For the beginner, a list of "don'ts" serve probably as useful advice rather than setting theoretically sound goals which may be difficult and frustrating to attain.

Firstly, it is worth remembering that low resistance electrodes (say 20–60 MΩ filled with HRP solution) make for better recording and current passing characteristics than high (fine tipped) resistance electrodes. A high voltage electrometer (Jochen *et al.*, 1981) does make using the latter type easier. But for standard electrodes and electrometers do not aim for too high a resistance. Do not make electrodes which seem to have beautiful tapers but which "wave around in the breeze". A rigid electrode is more likely to survive repeated tracking manoeuvres etc. Do not rely on resistance measurements made when choosing electrodes. By trial find the type of electrode which performs best *in situ*, and aim consistently to produce this type using shape and resistance only as guides to its suitability. Almost invariably the measured resistance increases when the electrode is inserted into the tissue of the preparation.

E. Keeping the Electrode Useable

It is essential always to make suitable "holes" in the pia overlying the central nervous system tissue, otherwise electrode damage or dimpling results. If the resistance increases above convenient levels and cannot be reduced by ringing or "backing off" (see following chapter) then its performance can be improved by further breaking the tip. Some workers (e.g. the Göteborg group—Lundberg, Jankowska, etc.) in fact favour controlled breakage rather than bevelling as a means of reducing intial resistance and forming a tip which penetrates tissue easily. In our experience, a bevelled electrode which exhibits high resistance during tracking can be improved easily by drawing the torn edge of a piece of lens tissue firmly across the tip. Don't be afraid of over-breaking the electrode—it can always be discarded as it would have been anyway in its plugged or blocked state. We find that electrodes which break to give resistances of about 20 MΩ *in situ* continue to give successful penetration and staining, with little tissue damage or HRP leakage.

4
Intrasomatic Recording and Horseradish Peroxidase Injection

I. CHOICE OF MICROELECTRODES

A recurrent topic of conversation between neurophysiologists is about the microelectrode characteristics required for particular tasks. Indeed, one sometimes feels that the craft of microelectrode manufacture has been elevated to the realms of wizardry. It is our contention that microelectrode manufacture and the choice of microelectrodes for use in different situations are both relatively straightforward. The other important point we wish to make is that, if you have made electrodes that perform satisfactorily in an experiment then, irrespective of whether they are "theoretically" good or bad (whatever that means) you should stick to using them. The ultimate test of a microelectrode is how it performs in the experiment.

We have discussed microelectrode manufacture in the previous chapter. Here we will try to offer some advice on how to choose microelectrodes for particular purposes, but always with the view to using the electrodes for both intracellular recording and HRP injection. This restriction immediately means that one has to compromise between an electrode that allows satisfactory intracellular recording and one that will permit the passage of sufficient electric current to inject the enzyme. For this reason, bevelling the tips of microelectrodes is very useful, as it increases the tip aperture without greatly affecting the outside diameter of the electrode at the tip.

There are two main variables in the anatomy of neuronal somata that determine the ease or difficulty with which they can be impaled by microelectrodes. These are their size and their shape. A third factor influencing intracellular recordings is the location of the neurones in the central nervous system. Mammalian neurones have somata that range in diameter by a factor of about 10, from below 10 μm for the smallest (for example in the substantia gelatinosa of the spinal cord, granule cells of the cerebellar cortex and various small cells in the cerebral cortex, olfactory bulb, etc.) to approaching 100 μm (for example dorsal root ganglion cells, α-motoneurones in the ventral horn of the spinal cord, the largest pyramidal cells in the cerebral cortex). As a generalization, it is true that the smaller the cell body then the smaller the electrode tip that is needed for impalement and the attainment of good recording conditions. It is surprising, however, what can be achieved with electrodes having tip sizes of even 1–2 μm, especially if these are bevelled or freshly broken and it is certainly incorrect to assume that because the target neurones are small one will need extrafine microelectrodes. In fact, we do not measure tip sizes of our microelectrodes, only their resistances *in situ*. When filled with 4–9% HRP in Tris-HCl buffer containing 0.2 M KCl, microelectrodes with resistances of 15 to 60 MΩ (immediately after bevelling and before inserting into the preparation) will cover the complete range of electrodes for impaling a variety of cells from the α-motoneurones to the small cells of substantia gelatinosa, as well as axons with diameters down to about 6 μm.

The shape of neuronal cell bodies affects the ease with which they can be impaled satisfactorily, especially for the smaller neurones. Above a certain size (probably upwards of about 25 μm) shape seems to have little effect, but below this size shape becomes more and more important. Thus, small spherical cell bodies, in which the nucleus occupies most of the available space and is surrounded by only a rim of cytoplasm, are the most difficult to impale. On the other hand, neurones with elongated cell bodies where one axis is much greater than the other (for example, 30 μm and 10 μm respectively) may allow stable intracellular recording and current passing for considerable lengths of time (up to half an hour or more).

In summary, the tip size of a microelectrode is seldom a determining factor in its ability to provide satisfactory intracellular recording and current injection. Tip sizes of the order of 1 μm, or more realistically electrode resistances of 15 to 60 MΩ (when the electrode is filled with 4–9% HRP in Tris-HCl buffer containing 0.2 M KCl) will prove equal to most tasks. The geometry of the tip may be more important than its size, as may be the geometry of the shank of the electrode.

In our experience, it is always worth bevelling a microelectrode tip. This provides a larger orifice, and lower resistance, for the same outside diameter. It is also often worth breaking an electrode tip, most easily by flicking it against a piece of tissue paper (or bumping against a glass microscope slide) if a series of bevelled electrodes provides little success. The point at which the tip breaks back depends on the geometry of the shank; a microelectrode with a long shank having almost parallel sides is difficult to break and usually finally breaks a long way back producing an electrode with too large a tip, a gradually tapering shank is generally the best shape to aim for. Again, as with most aspects of microelectrode technology, do experiment with different shapes of microelectrodes for different purposes. If high resistance electrodes must be used it is more satisfactory not to attempt bevelling, but to use a high voltage electrometer (see Chapter 2).

Finally, the location of the target neurones within the central nervous system may determine your choice of electrode. If the neurones are located within a few mm of the surface (say up to 3–4 mm) then electrode geometry is not usually too critical. In these situations, a fairly abrupt taper to the shank of the electrode can be tolerated and may even be advantageous for very superficial cells. If the target neurones are deep in the brain of a medium to large sized animal (rabbit, cat, monkey, sheep) then various problems occur. The microelectrode may frequently block as it is driven through the tissue, or the tip may deviate considerably from the intended direction. Both of these problems may be overcome by removing some brain overlying the target area, but this approach may not be feasible in your experiment. Again, do experiment with electrode design. However, it must be admitted that much patience will be required if your target neurones have to be approached through either the cerebral or cerebellar cortices.

II. IDENTIFICATION OF NEURONES

It has been a tenet of work done in the authors' laboratory that all neurones studied should be identified in terms of their inputs and outputs as well as their locations. This might be thought a great restriction, but the rewards of working with such neurones seem to us to far outweigh any restriction there might be. For neurones with long axons that project to known sites identification can be achieved by antidromic criteria. However, even short-axoned neurones can be identified on the basis of their input characteristics and, if they are favourably located (e.g. within a circumscribed nucleus), by their position.

A. Antidromic Identification

There is really only one satisfactory way to make a positive identification of a response as being antidromic and that is to use the collision test (Gordon and Miller, 1969), in which an orthodromic and an antidromic impulse are made to collide. The method has been discussed very clearly by Phillips and Porter (1977, pp. 54–57) and is illustrated in Fig. 4 (see also Fig. 5B). It is important to understand the time relations that allow a conclusion of antidromic activation to be made; sadly, many workers seem to be unclear about the period of unresponsiveness to antidromic stimulation. Following the discharge of an orthodromic action potential the unresponsive period must be equal to twice the conduction time plus the refractory period of the axon. The collision test may be used with both extracellular and intracellular recording conditions.

With intracellular recording it may be possible to use another criterion for antidromic invasion. If the impulse arises from a level of membrane potential that is below the normal orthodromic firing level and if there are no signs of synaptic depolarization preceding the impulse then it may be concluded that

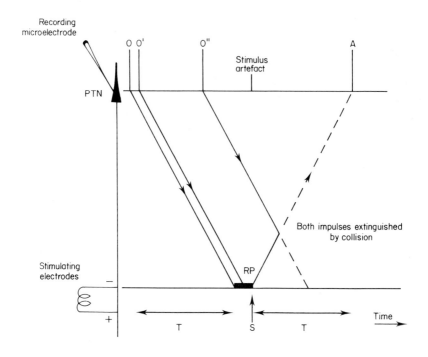

Fig. 4. Identification of an antidromic impulse by its collision with an orthodromic impulse. To the left of the figure is a diagrammatic representation of a neurone (PTN, pyramidal tract neurone) with a recording microelectrode in or near the cell body and stimulating electrodes at some distance away to stimulate the cell's axon. To the right of the "neurone" is a diagram explaining the time relations necessary to satisfy the collision criteria. Time is expressed along the horizontal lines at the top and bottom, the top line representing the recorded times of orthodromic (O, O', etc.) and antidromic (A) impulses. The antidromic impulse (A) is set up at time S at the site of the stimulating cathode. An orthodromic impulse arising at time O (or earlier) will arrive at the stimulating cathode before time S and will not prevent antidromic excitation. The orthodromic impulse O' will arrive at S such that it will make the axon refractory and prevent the setting up of an antidromic impulse. Orthodromic impulses arising at some later time (e.g. O'') will collide with, and extinguish, antidromic impulses set up at S. The minimum interval over which collision can occur is 2T + RP, i.e. twice the conduction time (T) between stimulating cathode and cell body (or recording position) plus the duration of the refractory period (RP). (From Phillips & Porter, 1977).

the impulse is antidromic (Fig. 5A). Some care is needed with this criterion, however, as it is possible to have a situation in which the supposedly antidromic activation excites an orthodromic input, where the excitatory postsynaptic potential has a very fast rise time and always leads to impulse initiation. In these circumstances, it may be impossible to differentiate antidromic invasion from orthodromic firing (Fig. 6). Furthermore, be aware of the possibility of failure of full antidromic invasion with only the initial segment (IS) or axon (M) spike remaining. To the unwary this might appear as an excitatory post-synaptic potential (Fig. 7). IS and M spikes will behave in an all-or-nothing manner, unlike EPSPs which should vary in amplitude with strength of stimulus.

The common criteria of antidromic firing, that is constant latency and high frequency following, are not reliable. There are always small variations in the latency of antidromic impulses, especially when it is cell bodies that are being invaded and, furthermore, some synaptically evoked impulses have remarkably constant latencies (Fig. 6). High frequency following is not a reliable indicator for antidromic firing (Fig. 6).

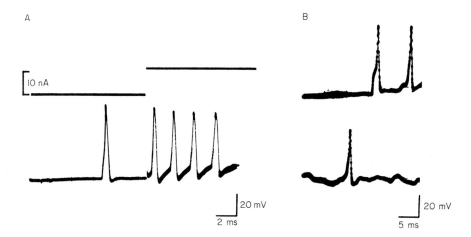

Fig. 5. Antidromic activation of neurones. (A) shows an antidromic impulse evoked in a spinocervical tract neurone, followed by a series of directly evoked impulses produced by passing a 12 nA current pulse through the intracellular electrode (upper trace). Note that the antidromic impulse arises from a flat base-line and shows a marked IS–SD separation on its rising phase. The directly evoked impulses arise from a lower (more depolarized) level of membrane potential and do not show such a clear IS–SD separation (10 superimposed traces). Note also the consistent latency of the antidromic spike. (B) shows collision of antidromic and orthodromic potentials. In the upper record an antidromic spike (spinocervical tract neurone) is seen, followed by an orthodromic spike. In the lower trace an orthodromic spike, produced by mechanical stimulation in the receptive field, occurs 5 ms before the expected antidromic spike and prevents it from being set up. Note the marked IS–SD separation in the antidromic spike.

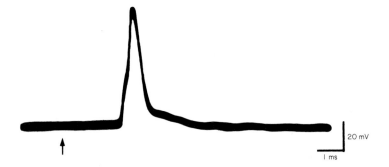

Fig. 6. False identification of antidromic action potential. Intracellular recording from a neurone presumed to be an α-motoneurone responding antidromically to stimulation of a muscle nerve. The impulse followed a stimulation rate of 200–300 Hz and appeared in all traces (10 superimposed on the photograph) at constant latency. There was no obvious EPSP preceding the spike, although the depolarization following it is presumably due to EPSP activity. Subsequent histological examination of this neurone showed it to be a ventral horn cell with an axon that ascended the lateral funiculus.

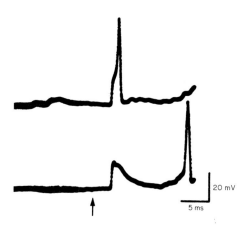

Fig. 7. Failure of SD component of antidromic impulse. The upper trace shows an antidromic impulse recorded from a lumbosacral spinocervical tract neurone (cat) in response to an electrical stimulus (arrow) to the cervical dorsolateral funiculus. Note the pronounced inflexion on the rising phase (IS–SD separation). The lower trace shows failure of the SD component of the antidromic spike and also shows that an orthodromic impulse, with full SD spike, is still capable of being generated by the late excitatory post-synaptic potential activity. Current had been passed into this neurone.

The placement of stimulating electrodes for antidromic identification is also important. If the axon terminates in a genuine "end location", such as the cerebral or cerebellar cortex, or leaves the central nervous system in ventral roots or cranial nerves, then stimulating at this "end location" or peripheral site is all that is required. Most axons, however, will end in a region of the central nervous system where stimulation within that region *per se* will not be sufficient for identification, since fibres of passage through the region will also be stimulated. In situations such as this, it is necessary to have two pairs of stimulating electrodes on either side of the target area, one pair either within the region or to the side of it in the direction of the cell under study and one pair of electrodes on the side of the region distal to the cell. An example of such an approach is shown in Fig. 8, which illustrates the way in which neurones of the spinocervical tract may be identified. In many experiments it will also be necessary to make one or more lesions in

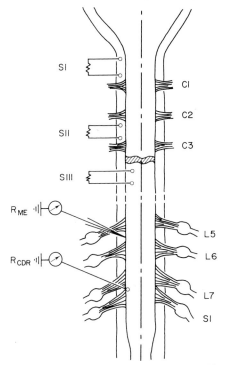

Fig. 8. Identification of spinocervical tract neurones in the cat. The tract terminates in the lateral cervical nucleus in the upper two cervical segments. Antidromic activation from below C2 (SII) and absence of such activation from rostral C1 (SI), or a greatly reduced axonal conduction velocity between the two stimulating sites, identifies spinocervical tract neurones recorded in the lumbosacral cord (R_{ME}). The dorsal columns are sectioned below SII to prevent orthodromic activation of the neurones via dorsal column fibres. R_{CDP} is a gross surface electrode for recording potentials from the cord dorsum.

the central nervous system to avoid the stimulation of pathways that excite the neurone orthodromically. Identification of spinocervical tract neurones, for example, is best performed with the dorsal columns sectioned between the proximal electrode pair and the recording site as shown in the Fig. 8. Refinements of these techniques will also allow the termination site of axons to be determined where a given nucleus or region projects to various targets. Thus, for corticospinal tract neurones, axonal terminations at different levels in the spinal cord may be determined by a series of stimulating electrodes along the length of the cord. Neurones with branched axons projecting to different areas can also be identified by using several pairs of stimulating electrodes.

B. Orthodromic Identification

Orthodromic identification may be used in conjunction with antidromic identification or by itself. When combined antidromic identification and orthodromic (input) data are used a very complete description of a neurone can be attained. When orthodromic methods alone are used the degree of identification varies with different neurone types. Orthodromic methods do, however, require a considerable knowledge of the inputs to the neurones under examination. An example of what can be achieved with such methods is the series of beautiful experiments carried out by Jankowska and her colleagues, which led to the identification of the Ia inhibitory interneurone in the spinal cord (Hultborn, *et al.* 1971a,b,c; Jankowska and Lindström, 1972; Jankowska and Roberts, 1972a,b).

Identification of short-axoned neurones excited by collaterals from neurones with long axons may be achieved relatively easily in many situations by using techniques appropriate for the *antidromic* identification of the long-axoned system. A classical example is the identification of Renshaw cells in the spinal cord by antidromic stimulation of motoneuronal axons in either the ventral roots or peripheral nerve (with dorsal roots cut).

If the neurone is known to be located within a circumscribed nucleus, this information may be used for identification purposes. Such identification will only be available if the neurone is stained and when the histological processing has been completed.

III. INTRACELLULAR PENETRATION

A. Microelectrode Tracking to Obtain Impalement

A systematic approach to intracellular penetration of a neurone has much to recommend it. The use of extracellular field potentials of a neurone allows systematic tracking, so that the microelectrode can be guided towards the cell.

This depends on being able to evoke a recognizable and consistent response from the neurone, the most useful response being an antidromic action potential uncontaminated with orthodromic impulses (or masking field potentials from other neurones), but this situation will only obtain in a minority of instances. If a recognizable response can be evoked, whatever its origin or character, then attempts to impale the neurone can be made in a systematic way. Neurones with no easily recognized responses, or neurones with little extracellular electrical field, cannot be tackled systematically. All one can do is aim the microelectrode in the appropriate direction and hope for the best; this sort of approach is required, for example, for the small neurones of substantia gelatinosa (lamina II) of the spinal cord.

When systematic tracking can be carried out, it is useful to have the microelectrode carried on a system allowing precise control of movements in three planes and also rotation about an axis within the two horizontal planes. In our system (see Chapter 2) the two horizontal planes are covered by carrying the stepping-motor drive (which drives the electrode in the vertical direction) on an arc mounted on a precision lathe movement giving control down to 10–20 μm. The arc can be rotated around an axis in the transverse plane of the animal and the microelectrode itself can be moved around the arc in one degree steps. In this way, the angle and position of approach of the electrode to its target can be accurately varied.

Once an appropriate target neurone is picked up in extracellular recordings, the microelectrode is moved systematically in attempts to increase the size of the extracellular signal.

B. Impaling the Cell

With good mechanical stability and the use of a stepping-motor drive, neurones may readily be penetrated and intracellular recordings made. The stepping motor drive is of great use, both to allow precise movements during the tracking procedure and also for cell penetration. During tracking the size of steps used varies according to the nature of the target cell. Neurones with large extracellular field potentials that can be recorded over long distances will allow quite large steps to be used during initial tracking, perhaps up to 20 μm steps. It is our practice to use steps smaller than this, even when tracking neurones with large extracellular fields, and we use steps of 8–10 μm routinely. Steps of 8–10 μm are also often excellent for penetration, especially for larger neurones. Some neurones, however, will require smaller steps as the electrode approaches them and certainly for penetration. We routinely use steps of 2 μm for smaller neurones (and often for axons, see Chapter 5). The main point about tracking and penetrating neurones, however, is to be flexible and to change the size of the steps to suit the circumstances.

Penetration of a neurone is signalled by the sudden appearance of the negative membrane potential and postsynaptic potentials. Action potentials should show overshoot to positive values. Often penetration is signalled by more ominous signs, such as a high frequency injury discharge and a membrane potential that decays rapidly to near zero values. The criteria for acceptable and stable impalement will be discussed later in this chapter. The first thing to do once impalement of a neurone has been achieved (and it seems as if the cell is in reasonable condition) is to check that the neurone that has been impaled is the same one that was being tracked.

C. Confirmation of Intracellular Identification

It is only too easy, in the excitement of obtaining a good intracellular recording, to overlook the important point that you might have impaled a neurone quite different to the one you have been tracking. Checks must be made to confirm that the target cell has been impaled.

In general, the neurone should respond in the same way as it did during extracellular tracking. Thus, any antidromic impulses should have the same latency and should arise from a level membrane potential with no signs of preceding EPSPs; also, orthodromic responses should have similar latencies and impulse patterns in both intracellular and extracellular conditions. There are some pitfalls to look out for, however: the view from inside a neurone is not always the same as the view from outside. For example, although receptive fields will often be the same under the two recording conditions this is not necessarily always the case. Neurones with subliminal fringes to their receptive fields may exhibit larger fields when penetrated by an electrode, especially if the penetration has resulted in a lower membrane potential (see Fig. 9). There are other phenomena that one needs to be aware of as well. Thus, an antidromic impulse may fail to invade the cell completely shortly after impalement or (more usually) may fail to invade partially so that only an M or IS spike is seen (see Fig. 7). The unwary might consider these potentials to represent synaptic events, but their all-or-nothing nature is easily confirmed.

If the microelectrode contains chloride ions (Cl⁻), as it will if the HRP is contained in a Tris-HCl buffer with additional KCl (as in our standard solution), then IPSPs that are initially hyperpolarizing events may quickly be reversed into depolarizing potentials. Again, this can trap the unwary; a stimulus that previously led to inhibition now appears to cause EPSPs and can lead to doubts as to whether the neurone tracked extracellularly has been impaled. This phenomenon also means that, if an aim of the experiment is to

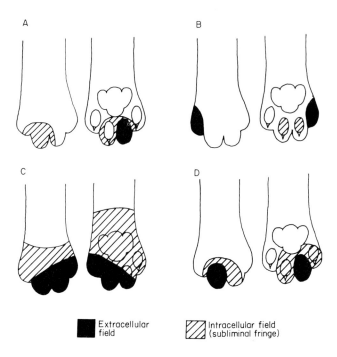

█ Extracellular field

▨ Intracellular field (subliminal fringe)

Fig. 9. Changes in receptive fields upon penetrating a neurone. (A, B, C and D) show cutaneous receptive fields of neurones of the post-synaptic dorsal column pathway in the cat. In each diagram the solid black area was the field outlined during extracellular recording and was that area from which impulses could be evoked. Upon penetrating the neurone with the microelectrode the receptive fields were shown to include the cross-hatched areas. Excitatory post-synaptic potentials could be evoked and then, within a matter of seconds, the impulse-firing zone expanded into all or part of the area.

examine post-synaptic potentials in addition to injecting HRP, then the use of chloride containing solutions has to be avoided.

Intracellular impalements may be made into areas of the neurone other than the cell body. Large dendrites or the cell's axon may be penetrated, and in certain experimental situations these may be the preferred sites. If the cell's dendrites are capable of carrying action potentials it may be difficult to know whether the electrode is in a dendrite or the cell body; if the recording conditions are good it probably will not matter anyway. In many cells, dendritic impalements are not maintained and the electrode "comes out" to reveal an extracellular field similar to (although usually of less amplitude than) that recorded before impalement. This is often the usual sign that a dendrite has been entered. Axonal impalements, if made far from the cell body, are recognized by the absence of post-synaptic potentials. Axonal impalements near the cell body, usually within a few mm of it, are signalled by the presence of low amplitude synaptic noise. If the aim of the experiment is to record from either dendrites or juxtasomatic axons, HRP injection may not be the most suitable marker for confirmation of such impalements, as it so easily stains the whole cell. Axonal impalements may be confirmed as the cell body and dendrites are often only lightly stained under these circumstances (Fig. 32) but dendritic impalements are less easily ascertained unless there is some HRP leakage around the site of electrode penetration. Procion Yellow, which remains localized near the injection site if the preparation is perfused very soon after injection, may be the marker of choice (see Llinás and Nicholson, 1971).

D. Criteria for Good Impalement

If a primary aim of the experiment is to study the physiological properties of neurones as well as to inject them with HRP, then it is imperative that the neurones are in as good a condition as possible: the microelectrode must produce little or no damage. There are standard criteria for good intracellular impalements that have been accepted and refined since the pioneering work on motoneurones from Eccles' laboratory. These criteria are as follows. (i) There should be a high, stable resting membrane potential. The magnitude of the potential should be as close to 70 mV as possible, although in practice values of over 40 mV are generally acceptable. (ii) Action potentials should overshoot the zero potential level. (iii) The microelectrode should not alter the firing rate of the neurone as recorded previous to impalement. This last criterion is the one that is most probably less strictly observed than the others. Most workers will be aware of the low frequency discharges that can be evoked by an intracellular microelectrode (as distinct from the short lasting—a second or so—high frequency burst that often ensues upon intial penetration), but it is also important to compare the cell's activity before and during impalement.

The above strict criteria are necessary for many experiments, especially when the neurone's membrane parameters are being studied. In other situations some relaxation can be tolerated. For example, a low level of discharge due to the microelectrode's presence may be of little significance when the effects of certain inputs are being studied. In such cases there will nearly always be a timed input and the neurone's response will be time-locked to it. It is also sometimes useful to allow the membrane potential to deteriorate so that hyperpolarizing IPSPs become larger (although EPSPs may be smaller) and deterioration to the point where the cell's impulse initiating mechanism fails may be advantageous for studying synaptic potentials without interference from action potentials (see Hongo et al., 1966).

From the point of view of intracellular staining good impalements are also generally desirable. Indeed, for the larger neurones we think they are mandatory, especially if the tissue is to be taken to the level of the electron microscope. Neurones with low membrane potentials during HRP injection often appear somewhat disorganized, even under the light microscope. In our experience, the best results are obtained not only when the neurone is in good condition before HRP injection, but also when it is in good condition at the end, i.e. it has a reasonably high membrane potential with overshooting action potentials. It has to be admitted, however, that these ideal circumstances are not always possible, especially when small neurones are being studied. In this latter situation, membrane potentials may only be a few tens of mV in size and the impulse generating mechanism may deteriorate very soon after penetration, and yet the subsequent histological appearance of the stained cell can be most satisfactory, even at the electron microscope level. Again, the experimenter has to be opportunist and take what comes.

IV. CURRENT INJECTION

A. Continuous Recording During Current Injection

It is very helpful if a continuous intracellular recording is made during the injection of HRP (or other intracellular marker). In this way, the state of the neurone can be assessed and, most importantly, injection can be terminated if the electrode comes out of the cell. This latter point is of prime significance for two reasons: first, in order to prevent extracellular deposition of HRP, which will lead to the staining of a group of cells close to the electrode tip (see Graybiel and Devor, 1974); and second, to ensure that only the neurone under study has been injected.

The most convenient way to monitor cell properties and inject HRP is to use a current pulse that is being passed only part of the time. The particular protocol we favour is to have the oscilloscope triggered every 600 ms and to pass a rectangular pulse of 450 ms duration that starts 10 ms–40 ms or so after the beginning of the oscilloscope sweep (see Fig. 10). This allows the first part of the sweep to be used for inspection of the cell's membrane potential and any evoked response. Sudden shifts in membrane potential, indicative of imminent loss of the intracellular recording, may be compensated in some instances by moving the microelectrode a few μm in either up or down directions. Deterioration of the condition of the neurone will also indicate when to stop current passing; sometimes, a period of rest will lead to recovery of impulse initiation and increase in membrane potential so that more current may be injected. Judgement as to when current injection should cease will depend on experience, but a sudden depolarization to values of the order of 20 mV or less usually suggests that more HRP injection will lead to extracellular spillage. Very small neurones, however, may be satisfactorily stained

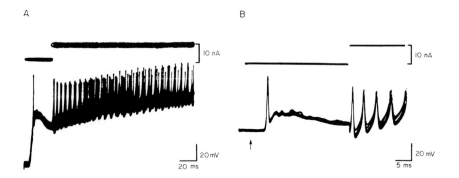

Fig. 10. Intracellular current passing. Both (A and B) are from the same cell (a post-synaptic dorsal column neurone), (A) at a slower time base than (B). In each figure, the upper trace is the current monitor, 8 nA in (A) and 10 nA in (B). The lower trace is the intracellular recording. Note the current-induced firing of the neurone in (A), lasting the duration of the trace (160 ms); in fact, the cell fired throughout the current pulse (450 ms), initially. In (B) the antidromic action potential, evoked from the cervical dorsal columns, can be clearly seen to arise from a flat base-line; it is followed by an excitatory post-synaptic potential complex. The current-evoked spikes are seen to arise from a ramp-like depolariza-tion. The intracellular record is quite well-balanced, particularly initially. (5 superimposed sweeps in each figure.)

by the passage of very little current when their membrane potentials are low.

During the current injection pulse, it is also useful to make recordings. An electrometer with bridge balancing facilities is necessary (see Chapter 2). With the record balanced (Fig. 5A), it is possible to ascertain whether the electrometer is capable of passing the applied current (blocking of the electrode). As saturation of the electrometer approaches it becomes increasingly difficult to balance the recording until no biological signals are recordable during part or all of the current pulse. By keeping the current injection pulse to amplitudes below saturation one can assume that at least some HRP should be passing out of the microelectrode.

B. The Injection Current

HRP can be injected into neurones by means of a depolarizing current or by pressure. We use current injection. As mentioned above, the protocol we favour is one where the current is being passed for 75% of the time. Rect-angular depolarizing currents of 450 ms duration, repeated every 600 ms, are used, and the current pulse is monitored on one beam of the oscilloscope (Figs 5A, 10). Advantages of this method include the ability to assess the state of the neurone between pulses together with the fact that pulses of current (rather than a steady DC current) help to prevent electrode blocking.

Shorter, higher frequency pulses may be used, but they restrict the recording of the cell's state. However, such short higher frequency pulses have been used successfully (e.g. Light and Perl, 1979). When high frequency (50–200 Hz), short (> 20 ms) pulses are used, it is necessary to stop the injection every few seconds to monitor the state of the impaled neurone.

Once an intracellular penetration of satisfactory quality has been achieved and it is decided to inject HRP, then the procedure is as follows. The current pulse is switched on and its amplitude slowly brought up from zero, with the state of the cell being monitored all the time and the electrometer being balanced. Any deterioration of the cell's condition should lead to cessation of current increase, or even a lowering of the current amplitude. The aim is to increase the current pulse to the largest value possible without cell deterioration or blocking of the electrometer, up to a value of about 20 nA. In our experience, there is no need to increase the amplitude of the injection current further than this.

The current pulse we use is a rectangular one, simply because of the ease of providing such a pulse from our stimulator. We have not tried any variations of the rise and fall times of the pulse. It is conceivable that a more slowly rising pulse might help in situations where a rectangular pulse causes problems of instability in the intracellular recording.

The amplitude of the injection current and the total amount of current passed in order to obtain satisfactory staining of neurones varies depending on the neurones under study. With regard to the amplitude of the current, the current passing capabilities of the microelectrode may also be a limiting factor. For large neurones and those smaller ones where stable intracellular penetrations are relatively easy to maintain it is advisable to use currents of 10 nA or greater in amplitude. It has been our experience that some sort of threshold effect for HRP injection appears to exist around 10 nA, in the sense that the same amount of current (expressed as nA.min) passed with larger amplitudes provides much superior staining in comparison with smaller amplitudes passed for a longer time. When studying very small cells, such as those of the substantia gelatinosa, we have had excellent results with small currents of 2–5 nA passed for only two to three minutes. The total amount of current needed for the larger cells is of the order of 75–100 nA.min, and it is advisable not to inject more than this, as one does not see, for example, greater lengths of axon branches stained, but rather disorganization of the cell and extracellular leakage of the HRP reaction product.

The depolarizing current pulse necessary to eject HRP from the microelectrode will excite the neurone. During the early phases of injection, at least, the cell will probably fire throughout the injection current pulse. In many neurones there will be deterioration of the spike generating mechanism at some time and the long train of current-evoked impulses will cease. There will still, however, be an intial one or two impulses, albeit of greatly reduced amplitude, evoked at the onset of the current pulse.

In addition to injecting HRP, the current pulse itself may be used for electrophysiological studies. One useful observation that is well worth making as a routine measurement is the size of the current necessary to fire the cell. With a satisfactory impalement this value should be of the order of 2–4 nA for the vast majority of mammalian neurones. Values higher than this indicate a poor impalement (a poor seal around the electrode tip) and suggest the microelectrode is pressing upon the cell membrane or is only partially inside the neurone. This can happen even when there is an apparent membrane potential of several tens of mV, especially if the membrane potential has arisen gradually rather than with a sudden step. The neurone's input resistance may be measured as may its firing rate to injected current.

C. Criteria for Success

Final assessment of the success (or lack of success) you have attained in staining neurones will only be possible when the material is examined under the microscope. It is possible, however, to make some predictions whether the injection is likely to have been successful during the electrophysiological experiment itself. In fact, it is usually easier to predict poor staining than good staining and it must be admitted that we are sometimes surprised by the results: predicted poor staining may be very good and occasionally predicted good staining may turn out to be very poor. There are some general criteria, however, that are usually fulfilled when neurones are satisfactorily stained.

Most important for success is the maintenance of good recording conditions throughout the period of current injection. The membrane potential should not drop suddenly and should be maintained at several tens of mV until withdrawal of the microelectrode. Neurones that maintain the ability to generate action potentials of considerable amplitude, preferably over-shooting the zero potential level, are nearly always well-stained. It is important that one finishes up injecting current into the same cell that one started with; again, the use of continuous recording is invaluable. Enough HRP has to be injected; use the information given in the preceding section and make sure that the electrode is in fact passing current (no blocking).

Even if the criteria given above are satisfied, there are still hazards ahead. These will be dealt with in the following chapters, but the perfusion of the tissue will need to be good and it will not be if the condition of the preparation deteriorates during the time between the injection of HRP into the neurone and the perfusion of the preparaion. If the condition of the preparation starts to deteriorate, it is best to cut your losses and perfuse as soon as possible.

V. PROBLEMS OF IMPALEMENT AND CURRENT INJECTION

A. Difficulties of Impalement

With adequate stability of the preparation, problems of impalement usually resolve into the choice of a suitable electrode configuration. Trial and error is necessary. There are some tricks that may be used, such as changing the size of the step in which the electrode is being moved, or tapping on the animal frame or the electrode holder. These last two resorts may be effective, but have not been in our hands. More effective has been the passing of very small (1–2 nA) pulses of depolarizing current through the electrode, if there are signs that the electrode is pressing on the cell membrane, i.e. very large extracellular spikes or "quasi-intracellular" ones associated with small DC shifts as the electrode is advanced. Sometimes impalements are achieved if the electrometer is caused to "ring" by increasing the capacity compensation.

B. Poor Intracellular Recordings

It is sometimes possible to improve the quality of an initially poor penetration by a variety of manoeuvres. If there are signs of damage, such as high frequency firing, then the passage of a hyperpolarizing current through the intracellular electrode may lead to cessation of firing and stabilize the penetration. We use the same pulse parameters as for current injection except, of course, the polarity of the pulse is reversed. It must be admitted, however, that, as often as not, the neurone recommences high frequency firing once the hyperpolarizing current pulses are discontinued.

If an initially good penetration deteriorates before you think an adequate amount of HRP has been injected, or before intracellular studies have been completed, it may be possible to improve the situation. Small steps of microelectrode movement (2 μm) can be effective in either the downwards or upwards direction. It is advisable to withdraw the electrode initially when trying this technique, otherwise it is possible to move the electrode right through the cell. If the cell deteriorates before any current has been passed, then the passage of current pulses, starting with a few nA, may be effective in rescuing the situation (in the same way that current passing may aid impalement).

C. Blocking of Microelectrode

This is one of the more frequent, and therefore frustrating, problems. Blocking while the electrode is in an extracellular position is usually only time wasting, but if it leads to the necessity to change the electrode then it can be more serious if a set of systematic search tracks are being made. In the extracellular position, an electrode may be cleared by withdrawing it to near the surface of the tissue and then causing the electrometer to "ring" several times. It is important to withdraw the electrode from close proximity to a neurone before attempting to clear it. Although "ringing" can aid impalement (see above), you are likely to finish up with an intracellular penetration with a blocked electrode that will not pass sufficient current for HRP injection. Alternatively, "ringing" may well lead to mortal injury to the neurone and for this reason is only used as an aid to impalement as a last resort. If "ringing" and withdrawal, repeated several times, fail to clear a blocked electrode then it may be withdrawn from the tissue completely and its tip cleaned, or more likely, broken back slightly with the aid of a piece of medical tissue—the tissue is dragged across the tip of the electrode. If you do carry out this operation, it is worth realizing that you may move the electrode slightly and any subsequent tracks cannot be guaranteed to be parallel to previous ones.

Once the microelectrode is in an intracellular position blocking becomes more serious, since it is much more difficult to clear an intracellular electrode. All you can do in this situation is to pass current pulses through the electrode (both positive and negative pulses). In our experience, it is rare for a satisfactory clearing of an intracellular blocked electrode to occur. For this reason, we advise that care is taken to keep the electrode clear during extracellular tracking by repeated "ringing" of the electrometer and also, if possible, by changing the electrode if it shows a propensity to block. Finally, in all situations use electrodes with as low a resistance as possible for the task in hand, preferably electrodes with bevelled or broken tips.

D. Under- and Over-staining, Cell Damage, etc.

These problems will obviously not be apparent until after the histological material has been examined, but in most instances the problem is at the stage of HRP injection and some attempts can be made to alleviate it. In general, experience will show how much injection current is needed for a particular neurone and, after a preliminary experiment or two, under-staining should not be due to the intentional passing of too little current for the job in hand, but rather to problems such as poor penetration, electrode blockage, etc. Over-staining and cell damage are more common problems, even in experienced hands. For most neurones of medium to large size, we now keep our injection current down to a total of about 100 nA.min if currents of 15–20 nA have been used. For small neurones, satisfactory injections can be achieved with only a few nA.min (e.g. 2–10). Again, experience will indicate what range of currents will be satisfactory in your particular experiment. If too much HRP is injected several problems arise. These include: a too dense staining reaction, leading to difficulty in sorting out crossing dendrites under the light microscope and complete inability to determine intracellular detail under the electron microscope; leakage of HRP from the cell, with consequent reaction product in the surrounding tissue and sometimes staining of nearby neurones; and distortion (swelling of cell body and dendrites) of the injected neurone. These problems, that appear at the histological level, will be discussed in more detail in later chapters.

VI. PHOTOGRAPHS AND MARKER ELECTRODES

In any successful experiment where several neurones (or axons) have been injected, perhaps in rather close proximity, problems may arise in identifying stained profiles under the microscope. Such problems are often more serious where a number of abortive attempts to stain neurones have been made, since in this situation there is a strong temptation to carry out further attempts near the site of the previous ones. This may well result in a number of stained neurones, with varying staining intensities, fairly close together.

In order to aid subsequent identification in the histological material, it is advisable to take detailed photographs of the brain or spinal cord surface and also to use marker electrodes. The surface photographs are used to plot the positions of electrode tracks where staining attempts have been made (or where electrodes have been broken). These photographs are also useful when tissue is being blocked for sectioning. Marker electrodes (microelectrode blanks) should be inserted to a known depth at a known distance away from tracks containing attempted injections. The blank electrodes are inserted using the same microdrive as is used for the recording microelectrode. Once inserted to the required depth, they can be cut with a pair of scissors so that only 1–2 mm protrudes above the surface of the brain or cord. These blanks should be removed before the tissue is sectioned (to prevent damage to the knife and/or the sections). They will leave easily observed tracks in the tissue.

5
Intra-axonal Recording and Horseradish Peroxidase Injection

I. CHOICE OF MICROELECTRODES

Intra-axonal recordings in mammalian preparations have been made for many years, some of the earliest being by Lundberg and co-workers from axons of the dorsal spinocerebellar tract in the cat (Laporte *et al.*, 1956). Recordings may be made from within either white or grey matter of the central nervous system, depending upon the problem under study. In general, it is easier to record from the white matter where axons are organized into tracts, but if the aim is to stain axons near their termination, so that their endings upon target neurones can be studied (and this is the usual aim), then HRP injections will often need to be made into the axon within grey matter.

The choice of microelectrodes for intra-axonal recording and current injection differs somewhat from that for intracellular recording. The general advice given for intracellular recording in the previous chapter will, however, be found to hold for intra-axonal electrodes. Again, the electrodes must be chosen for the job in hand and much empirical experimentation may be necessary. Usually, finer electrodes will be needed for intra-axonal penetrations. For axons with a diameter of about 6 μm and more, bevelled microelectrodes with initial resistances of the order of 20–50 MΩ will suffice. Axons of below 6 μm diameter are not easily impaled with bevelled electrodes and electrodes used straight from the puller (after filling) are best. These electrodes may have a very high resistance and do not allow easy current passage. A. R. Light (see Light and Perl, 1979) uses a high voltage electrometer. This lets him use electrodes of 120 MΩ to pass up to 20 nA of current. It is not easy to balance this type of electrode. Light also uses higher frequency, shorter current pulses for HRP injection than we use routinely.

In summary, it is in general more difficult to make satisfactory intra-axonal recordings and injections than intracellular (intrasomatic) ones, except where the axons are large and run in well-defined tracts, for example, in the dorsal spinocerebellar or spinocervical tracts. You will need to try many different types of microelectrodes before a satisfactory degree of success is achieved and the return from experiments may always be on the low side. If that turns out to be the case, then it is advantageous to back up your axonal studies with some other experiment that can be performed on the same preparation. This is something we have done in the past and does allow you to obtain some return from the investment of time, labour and expense.

II. IDENTIFICATION OF AXONS

As with studies on nerve cell bodies, it is advisable to identify axons in terms of their origin and projection. The same approach to identification, using both antidromic and orthodromic methods will be possible for axons. In addition, since the axon's cell body will probably have a known location, additional evidence for identification may be obtained by placing stimulating electrodes on either side of the soma location to excite the axon directly from the electrodes nearer to the recording site and to show failure of such excitation (but perhaps trans-synaptic excitation) from the electrodes further from the recording site.

With intra-axonal (and extra-axonal) recording it is impossible to determine whether an evoked action potential is travelling in the antidromic or orthodromic direction. It is only possible to tell whether the action potential has been set up by direct excitation of the axon or by indirect (trans-synaptic) excitation. With certain axons it will be possible, using other information about the organization of the system, to conclude whether a directly evoked action potential is travelling orthodromically or antidromically and the collision test can be applied. It is always necessary to bear in mind, however, that the axon under study may be projecting from "B" to "A" rather than the expected "A" to "B", even if a projection from "B" to "A" has not been described in the literature. Directly evoked impulses in axons will have a fixed latency, with less jitter than an antidromic impulse invading a nerve cell body and will usually be capable of following very high frequencies of stimulation (up to 1000 Hz) in a one to one fashion (Fig. 11). Again, there are possible exceptions: a directly evoked action potential, especially if travelling in the antidromic direction, may fail to invade beyond branch points, particularly at high frequencies.

There is a widespread belief among electrophysiologists that it is relatively straightforward to tell whether one is recording (extracellularly) from an axon or a cell body. It is our opinion that this belief is unsound. Although in certain circumstances an axonal extracellular action potential appears as a

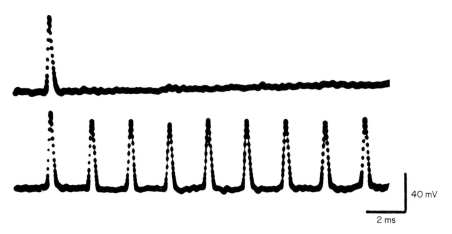

Fig. 11. Intra-axonal recording from a primary afferent fibre. The upper record is a single trace showing the axon giving a single impulse to electrical stimulation of the sciatic nerve (latency 1–6 ms). The lower trace shows the impulse following, in a one-to-one manner, a train of shocks at 500 Hz. Note the fixed latency of the response and the reduction in spike height after the first impulse. These records were taken using a digitizing oscilloscope.

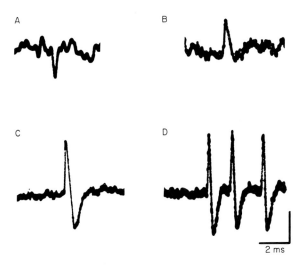

Fig. 12. Extracellular action potentials recorded from axons. (A, B and C) show some less common extracellularly recorded action potentials obtained from axons in the spinal cord. (A) a mono-phasic negative potential; (B) a monophasic positive; and (C) a bi-phasic positive–negative potential. (D) shows biphasic positive–negative potentials recorded from a nerve *cell* to illustrate the difficulties of making decisions about the source of extra-cellularly recorded potentials. Calibrations: 2 ms; 1 mV in (A), 100 mV in (B, C and D).

triphasic positive–negative–positive potential, usually when the electrode is far away from the axon, more often than not, the recorded potential is biphasic negative–positive, with the negative phase larger than the positive one. In these circumstances, there is a superficial resemblance to action potentials recorded extracellularly from cell bodies. Extra-axonally recorded potentials may also be monophasic negative (Fig. 12A) or positive (Fig. 12B) or biphasic positive–negative (Fig. 12C). It is commonly considered that an inflextion on the rising phase of an extracellularly recorded action potential is indicative (if not diagnostic) of a somatic source for the potential. We advise caution, as inflexions may be seen on the rising phase of an undoubted axonal recording. Very large extracellular action potentials may be recorded from both axons and cell bodies when the microelectrode is extremely close to the neurone. In these circumstances, the main component of the potential is positive and is followed by a small negativity. Axonal spikes recorded under these conditions can be several tens of mV in amplitude (as can the spikes recorded from cell bodies) and may or may not have an inflexion on their rising phases. These large extracellular potentials are, of course, recorded in the absence of any DC shift that would indicate intracellular penetration by the electrode.

III. INTRA-AXONAL PENETRATION

A. Impaling the Axon

Once the axon is impaled, there should, in most instances, be few doubts as to the axonal identification of the unit penetrated. In general, the lack of synaptic noise, with all action potentials arising from a flat resting potential, will confirm that the microelectrode is inside an axon (Figs 11, 13). There are still some pitfalls, however, for the unwary. An intra-axonal penetration made within a few mm of a nerve cell body will provide recordings in which synaptic potentials are evident (see for example, Gogan et al., 1983). The space constant of myelinated axons within the central nervous system is of the order of 2 mm (Barron and Matthews, 1938; Gogan et al., 1983), which means that electrotonic potentials fall to one half of their amplitude in 1.4 mm. Recordings made within a few mm of the cell body will display synaptic potentials, and more importantly perhaps an initial segment deflection on the rising phase of an orthodromically evoked potential if the potential arises there. Not only may such recordings lead to a mistaken belief that the electrode is inside a cell body, with subsequent abandonment of that particular unit, but the reverse problem may occur. That is, in an experiment in which the targets are cell bodies an axonal recording made close to a cell body may be accepted and HRP injected. In such instances, the cell may not be stained at all, or may be only very lightly stained (Fig. 32).

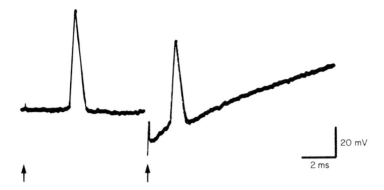

Fig. 13. Intra-axonal recording from a primary afferent fibre. Following peripheral nerve stimulation (first axon) there is an orthodromic action potential. At the second arrow a 0.5 nA depolarizing current pulse (450 ms duration) commenced and led to direct excitation of the axon at the rather long latency of about 1.0 ms. This recording is not well-balanced.

Impaling axons is a more chancy affair than impaling cells, in the sense that axons cannot be tracked systematically. Usually one picks up the first extracellular potentials from an axon within about 10–30 μm of the point at which impalement takes place. Sometimes an extra-axonal recording can be followed for further than this, especially when the axon has a large diameter. For example, Ia afferent fibres from muscle spindle primary endings in the cat with diameters of the order of 18–20 μm may be recorded over distances of 50 μm or more.

In our experience, it is easier to impale axons with relatively large electrode movements, even though the distance over which an axon may be recorded is often rather short. Thus, we use steps of the order of 8–10 μm for impaling axons. It must be admitted, however, that most of our experience is with the larger myelinated axons. Impalement of the smaller axons will probably necessitate smaller microelectrode movements.

B. Criteria for Good Impalement

Criteria for good intra-axonal impalements are similar to those for good intra-somatic impalements. Thus, a high resting potential of at least 40 mV with overshooting action potentials are required at initial penetration. The resting potential should be steady and not show fluctuations, (with respiratory movements, for example), or loss of the axon is imminent. A further test for a "robust" axonal impalement is to measure the threshold for initiating a directly evoked impulse by passing current through the intra-axonal electrode. This threshold should be of the order of 1 nA or (usually) less. (Fig. 13).

IV. CURRENT INJECTION

Details of current injection into single axons are essentially the same as those for injection into neuronal somata. There are a few points to bear in mind, however. First, try to inject as much HRP as possible. Good axonal staining is more difficult to achieve than good soma-dendritic staining. The HRP moves in both directions from the point of injection and often only one direction is of interest. The total lengths of stained axon required can be very great, as usually the aim is to stain an axon's terminal field (or several terminal fields) together with the synaptic boutons it carries. The more HRP injected the greater the chance of realizing this aim. The injection should, therefore, be carried out with as large a current pulse as the electrode will pass without signs of block or loss of the axon. A pulse amplitude of 20 nA, or greater if possible, should be attempted. In addition, pass the current for as long as possible, that is, for the whole time the axon remains in reasonable condition. In our experience, there is no risk of distortion with axons, as there is with cell bodies, if a lot of current is passed. Second, monitor the state of the intra-axonal penetration carefully throughout. If the impalement is made in an axon running in a tract, it is not uncommon to find spillage around the site of a poor impalement and other axons having taken up the HRP. The degree of staining shown by axons around such a spillage can be quite remarkable and this makes the subsequent histological analysis difficult or impossible.

V. PROBLEMS

A. Difficulties of Impalement

As for intra-somatic impalements, any problems are usually caused by the microelectrode. Undoubtedly, the tolerance in the microelectrode tip geometry is much less for axonal impalements. We cannot over-emphasize the need to keep persevering with different electrode variables if difficulty is encountered. Tapping the animal frame etc., passing small positive currents through the electrode or making the electrometer "ring" are all much less successful in aiding intra-axonal penetrations than they are for intra-somatic impalements.

B. Poor Intra-Axonal Recordings

It is less easy to improve a poor intra-axonal impalement than a poor intra-somatic one. It is occasionally possible to improve matters by very small movements of the electrode or by passing small current pulses. In our experience, however, these occasions are very rare indeed.

C. Blocking of the Microelectrode

This problem is, unfortunately, more common with the higher resistance microelectrodes necessary for intra-axonal recording. There is very little you can do to improve the situation. Occasionally passing a hyperpolarizing current leads to temporary clearing of the electrode.

D. Under- and Over-staining

Under-staining is common with intra-axonal injections. Try to pass as large a current as possible for as long a time as possible. Over-staining has never been a problem in our laboratory. There is, however, a related problem. This is the leakage of HRP around the site of electrode penetration. It arises when the impalement is of poor quality and can lead to considerable difficulty in analysing the histological material, as one or more adjacent axons may take up the enzyme and stain strongly. It can be overcome by continuous monitoring of the intra-axonal recording and cessation of injection if there are signs of the electrode coming out of the axon. In this latter context, beware of a gradual loss of resting potential—electrodes do not necessarily jump out of axons.

6
Histology

I. INTRODUCTION

The major emphasis in intracellular staining studies is the attempt to correlate the physiology of stained neurones with their light and electron microscopical appearance. The ability to take the material to the ultrastructural level opens up numerous possibilities that light microscopy alone is incapable of providing. Thus, detailed information becomes available on the ultrastructure of the input to stained neurones, as well as information on the ultrastructure of the stained neurones themselves. In addition, the combination of HRP injection with other techniques, such as Golgi staining, autoradiographic methods and immunocytochemical techniques, opens up a vast spectrum of possibilities for understanding the microcircuitry and transmitter candidates in small groups of neurones. For these reasons, it seems sensible to aim for a general set of histological procedures that will allow both light and electron microscopical examination of stained cells. Over the years, we have modified our original techniques of fixation and histochemical processing to allow material to be taken to this level of analysis.

Jankowska et al. (1976) were the first to attain the triple level of analysis: electrophysiology, light microscopy (LM) and electron microscopy (EM) of HRP injected neurones. Other groups have since developed similar methods (e.g. Cullheim and Kellerth, 1976; Gobel et al. 1980; Light et al., 1981; Réthelyi et al., 1982; Maxwell et al., 1982). In this Chapter we will describe our methods for fixation (perfusion), section cutting, HRP processing by both the diaminobenzidine (DAB) and pyrocatechol/p-phenylenediamine (PC-PPD) methods, section scanning and EM processing of stained sections, together with the counterstaining of HRP-stained material.

II. FIXATION BY PERFUSION

A. Apparatus Requirements and Solutions

The perfusion apparatus we now use is shown in Fig. 14. It was made by Mr R. Clark in our department. It consists of three flasks: two of 1 litre capacity are contained within a water bath at 38°C, the other flask is of 2 litre capacity and is not placed in the water bath. A pump and reducer valves allow the pressures in each flask to be controlled over a range of 0–500 mmHg. The water bath, flasks and pump etc. are mounted on a trolley, at the front of which is a removable stainless steel tray. This tray, on which the perfusion of the animal is carried out, is sloped and has drainage holes for removal of fluids. In the base of the trolley there is a tank into which the drained fluids are collected. The only other apparatus requirements are lengths of polythene tubing, a three-way valve, stopcocks and a large-bore stainless steel needle.

This apparatus allows accurate control of the temperature and pressure of the perfusion fluids. A reasonable assessment of the volumes used can also be obtained. We have found that the quality of fixation (at the EM level) produced by perfusion with this apparatus is greatly superior to other methods we have used.

Fig. 14. Perfusion apparatus used by the authors. For further descriptions see the text. The apparatus was made by R. Clark in our department.

The solutions used for perfusion (see Table 2) are as follows: (1) 0.9 %
NaCl (w/v) containing 0.1 % $NaNO_3$ and 10 units of heparin per ml; (2)
3 % paraformaldehyde, 3 % glutaraldehyde solution containing Na cacodylate
buffer at pH 7.4. Solution 1 (the $NaCl/NaNO_3$) is prepared as a stock solution
and the heparin is added immediately before it is placed in the flask, about
30 minutes before the perfusion. Solution 2 is made up on the day of the
experiment (about 3 litres for a cat) and stored in the refrigerator. One litre of
this fixative is placed in one of the warmed flasks about 30 min before perfusion and rather more than a litre is placed in the unwarmed flask about 4 min
before perfusion. Solution 2 is our standard fixative for combined light and
electron microscopical studies. If only light microscopy is to be performed it
is advisable to reduce the amount of glutaraldehyde, to about 1–1.5 %, as
glutaraldehyde reduces both the penetration of the fixed tissue by the histochemical reagents and also the sensitivity of the various histochemical methods
for the demonstration of HRP.

Table 2. Solutions for perfusion/fixation

Saline (0.9 %)
(i) 45 g NaCl in 5 litres distilled water
(ii) Add 400 units heparin per 500 ml saline
(iii) Add 0.4 ml of 1 % $NaNO_3$ solution per 500 ml saline

Fixative
(i) 240 ml glutaraldehyde plus 60 g paraformaldehyde (this gives 3 % of each aldehyde)
(ii) Add 667 ml 0.3 M Na cacodylate buffer and 1 M HCl to give pH of 7.4
(iii) Make up to 2 litres with distilled water

B. The Perfusion

The perfusion apparatus is prepared by filling the flasks contained in the
water bath with the $NaCl/NaNO_3$ and the fixative solutions (in separate
flasks!) and turning on the heating element. Once the solutions are at body
temperature, the cold fixative is placed in the third flask and the perfusion can
commence; this is from one to twelve hours after the intracellular injections
have been performed. Leaving the preparation for more than about 12 hours
following an injection leads to fainter staining, presumably due to extrusion
of the HRP from the cell, or to transport away from the site of injection.
Longer post-injection times do not lead to greater lengths of axon being
stained.

A few minutes before perfusing the preparation, 5000–10 000 units of
heparin are given i.v. The deeply anaesthetized (or decerebrate) animal is

then transferred to the perfusion tray on the front of the perfusion trolley, care being taken not to dislodge any marker electrodes left in the central nervous system. The animal is clamped on the perfusion tray in a simple head holder so that it is lying on the ventral surface of its trunk. In this position there is little risk of dislodging marker electrodes or of damaging exposed parts of the brain or spinal cord. It is, of course, possible to perfuse the animal when it is still in the stereotaxic head holder and/or spinal frame. We carried out the perfusion in this way for a number of years, but access is rather limited when the animal is in a frame and one also runs the risk of damaging apparatus, particularly the head holder and the frame, by the very strong fixatives. Furthermore, with the animal on the experimental table there is inevitably a lot more spillage of fixatives than when a custom built perfusion system is used and it is certainly advisable to try to avoid spillage of aldehydes, and their inhalation.

For spinal cord fixation we perfuse via the aorta, in the caudal direction for the lumbosacral cord and in either the caudal or rostral direction if the cervical cord is also required. For brain we have found that perfusing through the aorta in the rostral direction will also provide good fixation. Obviously different workers will use whatever methods give them the required fixation. Thus, Light and Perl (see Metz *et al.*, 1982) perfuse through the carotid cannula they have previously used to measure arterial blood pressure.

With the animal in the prone position, the left side of the thoracic cavity is opened widely by removing parts of several ribs and the exposure is held open with retractors to allow easy access to the descending aorta. The pericardium is held with artery forceps and pulled laterally to allow access to the right atrium which is incised to allow venous drainage and the large bore needle is inserted immediately into the aorta and secured by a ligature (after having previously flushed the tubing of the perfusion system to remove all air bubbles).

The perfusion is begun with 200–500 ml of the warm $NaCl/NaNO_3$ solution at 150–200 mmHg. Surface blood vessels on the spinal cord or brain should clear rapidly. This saline rinse is followed immediately by one litre of warm fixative at the same pressure and again immediately by one litre of the cold fixative at pressures of 150 mmHg, reducing to about 100 mmHg. These volumes are suitable for animals the size of a cat; appropriate adjustments should be made for larger or smaller animals. The whole perfusion process should take less than ten minutes. A good perfusion is essential, not only to fix the tissue but also to remove red blood cells and other vascular elements that contain endogenous peroxidase that might stain during the histochemistry.

Following perfusion, the parts of the brain or spinal cord that are of interest are carefully removed and rinsed in 0.1 M phosphate buffer, pH 7.4 (Table 3).

Using the marker electrodes and surface landmarks, such as blood vessels, sulci or segmental boundaries as guides (we find photographs of the brain or cord *in situ* more useful than drawings), the tissue is blocked into suitable pieces (e.g. lengths of spinal cord about 5–6 mm long). We usually slice one corner off the block or slice off one sliver of tissue from the ventrolateral quadrant of the spinal cord to aid orientation during subsequent serial sectioning and section mounting. The marker electrodes can now be carefully removed (they will have left their tracks in the tissue) and it is useful at this stage to remove all traces of the pia-arachnoid membranes. In any event, these membranes will need to be removed before sectioning is started. Any neural attachments, such as spinal roots or cranial nerves, should be cut off close to the central nervous system. The blocks of tissue are then stored overnight in the fixative solution in the refrigerator.

III. HISTOLOGICAL PROCEDURES

A. Cutting Sections

For light microscopy, sections may be cut on the freezing microtome. For combined light and electron microscopy (or Golgi) sections should be cut using a tissue slicer, such as the Vibratome (Oxford Laboratories, 1149 Chess Drive, Foster City, California 94907) or the Vibroslice (Campden Instruments Ltd., 186 Campden Hill Road, London).

1. Light microscopy
If the tissue is to be examined only by light microscopy, then cutting sections with a freezing microtome is the easiest. In order to prevent ice crystals forming in the sections, some cryoprotection is necessary. Soaking the blocks of tissue in 30% sucrose in 0.1 M phosphate buffer (pH 7.4) overnight or for two to four hours will prevent ice crystal formation.

The thickness at which sections are cut for light microscopy will be determined, in large part, by the nature of the stained material. A compromise is usually necessary between cutting thin sections that allow considerable detail to be seen under high power (oil immersion or interference optics) on the one hand and thick sections that reduce the labour of reconstructing the stained neurones to provide three dimensional data. A further factor is the size of the stained neurones: thus very small neurones with locally ramifying dendrites and axons will permit the use of thin sections, whereas large neurones, and especially stained axons that can be followed for many mm, will demand thick sections in order to keep the task of reconstruction within reasonable limits. In our experience, sections of between 50 and 100 μm thickness will satisfy the demands of most material, the 50 μm sections being used for material containing smaller neurones and the 100 μm sections for material containing large neurones and axons that can be followed for long distances. However, we have satisfactorily handled much thicker slices of tissue than this, of the order of 1–2 mm, and these slices may be processed with Golgi methods (see below).

2. Combined light and electron microscopy

There is much to be said for using a combined method. If electron microscope facilities are available then interesting material can be taken to this level of analysis. The tissue blocks are attached to the Vibratome or Vibroslice base plate with cyano-acrylate glue. A Plasticene wall is constructed around the block and a 4% solution of agar (melted and then cooled to 40°C) is poured over and around the block of tissue. When the agar has set, the Plasticene is removed leaving the block encapsulated in the supporting agar medium. Sections are cut on the Vibratome or Vibroslice at a thickness of 50–100 μm.

B. Collecting Serial Sections

In order to allow three dimensional reconstruction of the stained material, serial sections are required and no loss of material should be tolerated. The sections are collected serially in trays (Fig. 15). The trays, which are home-made, consist of rigid Perspex sheets 10 mm thick. Holes of 10 mm diameter

Fig. 15. Tray with chambers for collecting serial sections. For further descriptions see the text.

are drilled in a regular array and one side of the sheet is covered by stainless steel mesh. The mesh is attached to the Perspex by melting the Perspex using chloroform. The sections are collected serially into the compartments of the tray and remain in their compartments during the histochemical processing. This avoids handling the sections and the risk of losing or damaging them.

The collecting trays are immersed in 0.1 M phosphate buffer, pH 7.4 for light microscopy, or 0.15 M cacodylate buffer, pH 7.4 for electron microscopy (Table 3). Care must be taken to keep the depth of fluid in which the trays are placed at below the level that would cause the sections to float out of their compartments. The collecting trays are fitted with feet that keep them away from the bottom of the containers holding the buffer (and later the containers holding other reagents), so that the buffer can enter the tray's compartments through the mesh.

Table 3. Buffered solutions

Phosphate buffer (0.1 M)
(i)   Solution A: 42.6 g Na_2HPO_4 in 3 litres distilled water
(ii) Solution B: 31.2 g NaH_2PO_4 in 2 litres distilled water
(iii) Add solution B to solution A until pH is 7.6

Na cacodylate buffer (0.3 M)
(i)   65.8 g Na cacodylate
(ii) 13.8 ml 1 M HCl
(iii) Make up to 1 litre with distilled water
(N.B. use at 0.15 M concentration for rinsing sections)

C. Histochemical Methods for the Demonstration of HRP

1. Introduction

With H_2O_2 as substrate, HRP catalyses the reduction of some compounds (chromogens) to form a visible precipitant or reaction product. Graham and Karnovsky (1966) originally used diaminobenzidine (DAB) as the chromogen and this continues to be used by some workers. It has, however, been superceded by other chromogens, especially when HRP is used for retrograde or anterograde pathway tracing. The search for and development of alternative histochemical procedures has been prompted by the desire to find safer chromogens (some are carcinogenic) and to find chromogens with greater sensitivity.

In view of the carcinogenicity and toxicity of many of the agents used in HRP histochemistry, and in standard histological processing for light and electron microscopy (e.g. aldehydes, cacodylate, osmic acid, epoxy resins), we wish to stress that full safety precautions should be used when potentially harmful substances are handled. We carry out histochemical reactions and much EM processing in a fume cupboard, and wear protective clothing and face masks at all times when there is a risk.

When HRP is injected into single neurones through a microelectrode, so much HRP gets in that sensitivity of the chromogen is not a limiting factor. The most important criteria are that the chromogen should not produce artefacts and should be sensitive in a histochemical reaction that leaves the ultrastructure of the nervous system in an acceptable state. These two criteria are met by DAB and the PC-PPD chromogen (see below), both of which are sensitive enough for work involving intracellular injection. Some additional sensitivity can be achieved if the reaction product is intensified by cobalt precipitation (Adams, 1977). The use of tetramethylbenzidine as a chromogen (Mesalum, 1978) provides much greater sensitivity and in many situations is the method of choice for retrograde and anterograde pathway tracing experiments where it can produce a Golgi-like staining of neurones. For intracellular injection studies, however, TMB has not been particularly successful, at least certainly not in our hands.

2. The 3,3'-diaminobenzidine (DAB) method

The original DAB technique (Graham and Karnovsky, 1966) has been successively modified by many workers for use on the nervous system (e.g. LaVail and LaVail, 1972; Graybiel and Devor, 1974; Snow et al., 1976; Jankowska et al., 1976; Light and Durkovic, 1976) and DAB was the chromogen used in the first successful intracellular staining experiments. DAB methodology has been fully described and discussed by Warr et al. (1981). The method as originally used in our laboratory was described by Snow et al. (1976) and the earliest material from those experiments has retained labelling details since that time (May 1975) and fine distal dendritic branches of stained neurones can still be resolved in the light microscope. It should be noted, however, that fine collateral branches of axons have faded somewhat.

Unfortunately, the carcinogenic properties of DAB make handling it hazardous. We ceased using it routinely in 1977 (when the PC-PPD method became available). DAB processing is still the method used in some laboratories to demonstrate intracellularly stained neurones, and we use it occasionally. It has an advantage for ultrastructural studies in that the fainter staining intensity (less electron density) often allows more intracellular detail to be observed than when PC-PPD is used.

The DAB procedure (for details of the procedure, using the safe ISOPAC of DAB available from Sigma London Chemical Co., Poole, Dorset, see Table 4) is as follows:

(i) The serial sections are collected in trays whose compartments have a stainless steel mesh on the bottom (see pp. 60–61). The trays are immersed in 0.1 M phosphate buffer at pH 7.4 or 0.15 M cacodylate buffer at pH 7.4 (for EM) (Table 3).

(ii) The sections are now pre-incubated in DAB by transferring them, in the trays, to a solution of 0.1 M phosphate or 0.15 M cacodylate buffer (pH 7.4) containing 0.05% DAB for 15–20 min at room temperature. The trays should be gently shaken. Note that this stage may be omitted, see Table 4.

(iii) The sections are washed by passing the trays through two to three changes of 0.1 M phosphate or 0.15 M cacodylate buffer (pH 7.4) with gentle shaking.

(iv) The main incubation step now follows. The trays are transferred to a 0.1 M phosphate or 0.15 cacodylate buffer solution (pH 7.4) containing 0.05% DAB and one to two drops of 30% H_2O_2/100 ml solution. The trays are gently shaken for 15–20 min. This reaction may be carried out at room temperatures above about 17°C. If the room temperature is below this value, we carry out the incubation at 25°C in an oven.

(v) The sections are now washed in three changes of the buffer.

(vi) Depending on whether or not the material is to be further processed for electron microscopy, the procedure varies. Preparation for EM is given in detail below. For light microscopy the sections are now mounted serially on gelatin-coated glass slides and left to dry in air for two to four hours. Then they are placed in formalin vapour overnight by sitting the slide holders under a bell jar or inverted glass trough containing a Petri dish full of a stock (3%) solution of formalin.

(vii) The mounted sections are dehydrated in an ascending series of alcohols (50% for 5 min; 70% for 10 min; 90% for 10 min; 100% for 10 min; 100% for a further 10 min and cleared in xylene (5 min and 10 min). Finally they are coverslipped using UV-inert mounting medium (Gurr).

Table 4. The DAB method (using ISOPAC—100 mg in 100 ml rubber scaled bottle)

(i) Inject 100 ml Tris-HCl buffer pH 7.6
(ii) Inject 0.4 ml of 20% H_2O_2
(iii) Shake to dissolve
(iv) Remove solution from bottle and place in 200 ml beaker and add further 100 ml Tris-HCl buffer
(v) React sections for 30 min
(vi) Rinse in phosphate or cacodylate buffer (twice)

3. The pyrocatechol/p-phenylenediamine (PC-PPD) method

This method is due to Hanker *et al.* (1977), who introduced a mixture of pyrocatechol and *p*-phenylenediamine as substrates for the HRP reaction. The compounds were suggested to be non-carcinogenic, but we would still advise extreme caution in handling them and that all stages of the procedure be carried out with the same safety precautions used for DAB. The sensitivity of the method is excellent for intracellular staining and is now our routine method. With good perfusion of the material the background staining is low (but not as good as with DAB), non-specific peroxidases are not as densely stained as with DAB and staining of dendrites, axons and axon collaterals is dense. Indeed, for EM studies the staining intensity is sometimes too great to allow ultrastructural details of the injected neurones to be studied in detail. When less HRP is injected, by using less total injection current or reducing the incubation time of the HRP reaction, excellent detail of the ultrastructure of injected neurones may be obtained.

Table 5. The PC-PPD method

(i)   225 mg PC-PPD reagent
(ii) Add to 150 ml buffer (0.1 M phosphate, 0.15 M cacodylate or 0.3 M Tris-HCl)
(iii) Add 0.2 ml 30% H_2O_2
(iv) Mix until ingredients dissolved
(v)   React for 15–20 min, then rinse sections in buffer
(vi) Rinse in phosphate or cacodylate buffer (twice)

Our procedure for the PC-PPD reaction (for details of the solutions used see Table 5) is as follows:

(i) The serial sections are collected in 0.1 M phosphate or 0.15 M cacodylate buffer (pH 7.4) as for the DAB reaction.

(ii) The sections are pre-incubated in 0.15% PC-PPD reagent (available ready mixed as "Hanker-Yates" reagent from Sigma) in buffer (pH 7.4) for 15–20 min with gentle shaking at room temperature. This step may be omitted.

(iii) The sections are passed through two to three changes of buffer, with gentle shaking.

(iv) The incubation step follows. The sections are transferred to a solution of buffer (pH 7.4) containing 0.15% PC-PPD reagent and one to two drops of 30% HRP per 100 ml solution. The sections are gently shaken in this solution for 15–20 min at room temperature (or at 25°C if the room temperature is below about 17°C).

(v) The sections are washed in three changes of buffer, for about five minutes in each.

(vi) and (vii) Steps (vi) and (vii) are the same as for the DAB method, if the material is for light microscopy only.

4. Variations on the basic procedures

(a) *Intensification with cobalt* Adams (1977) described a method for the intensification of the HRP reaction product in histochemical processes. We now use this method on much of our material. It is particularly effective in improving the staining of fine axonal branches and terminal boutons. The protocol is similar with either DAB or PC-PPD mixture as the chromogen.

The intensification is carried out at the pre-incubation stage (step (ii) in the above). The sections are placed in a solution (Table 6) of 0.1 M cacodylate buffer (pH 5.1) containing 0.6% cobalt chloride, 0.4% nickel sulphate and either 0.05% DAB or 0.15% PC-PPD mixture. The sections are gently agitated in this solution for 15–20 min at room temperature.

Table 6. Cobalt intensification of PC-PPD reaction

A. Pre-incubation for 15–20 min in:
  (i)   225 ml of 0.1 M Na cacodylate buffer pH 5.1
  (ii) 0.338 g PC-PPD reagent
  (iii) 1.35 g cobalt chloride
  (iv) 0.9 g nickel sulphate

B. Rinse in two changes phosphate buffer, pH 7.4

C. Incubate for 15–20 min in:
  (i)   225 ml of 0.1 M Na cacodylate buffer, pH 5.1
  (ii) 0.338 g PC-PPD reagent
  (ii) Four drops of 30% H_2O_2

(b) *Counterstaining* It is sometimes very useful to counterstain light microscope sections, to delineate cytoarchitectural areas or to define possible groups of target neurones, for example. We do not counterstain routinely at the histochemical processing stage, but after having examined, drawn and photographed the material in unstained sections. If counterstaining is required, the coverslips are removed by immersion in xylene for two to three days. The sections are then rehydrated through a descending series of alcohols, counterstained with a Nissl stain, such as cresyl violet or methylene green (we find the latter especially useful), and then dehydrated and coverslipped as described above. We have noted that the rehydration-counterstaining procedure actually enhances the contrast between HRP-stained profiles and the background tissue. Tissue counterstained with methylene green is shown in Fig. 36.

(c) *Rapid formalin method* For those workers who are extremely impatient to see their material at the light microscope level (and we include ourselves under this heading) the rapid formalin method allows mounted and coverslipped sections to be examined on the afternoon following the day of intracellular injection. The procedure is the same as described above except that at step (vi) the slides are placed under a bell jar containing a Petri dish full of formalin and the set-up is warmed to 60°C for about 30 min. This rapid method is very useful if time is short, for example, towards the end of the week, or when one needs to know the results from one experiment before starting the next one.

(*d*) *Combination with Golgi methods* It is perfectly feasible to combine intracellular HRP staining with rapid Golgi methods. For this purpose, thick (1 mm) sections are cut, on the Vibratome or Vibroslice, and processed for HRP as described above, except that both the pre-incubation and incubation stages, steps (ii) and (iv), are prolonged to about one hour each. After the main incubation step, the thick sections are kept in 0.1 M phosphate buffer (pH 7.4) until the rapid Golgi processing is carried out. Once successfully impregnated, the thick sections can be re-embedded in agar and cut at 50–100 μm, on the Vibratome or Vibroslice, for detailed light microscopy and further processed for electron microscopy.

5. *Some general points on the histochemical procedure*
 (i) All used solutions and utensils should be deactivated by rinsing in an industrial bleach solution such as "Chloros" (5% sodium hypochlorite). Utensils should then be thoroughly rinsed several times in cold running water. It is essential to remove all traces of bleach from the reaction trays, section holders, etc. to prevent accidental deactivation of any subsequent histochemical reactions.
 (ii) All reactant solutions should be prepared less than one hour before use.
 (iii) All solutions should be filtered before use.

D. Preparation of Material for Electron Microscopical Analysis

The following methods have been developed in our laboratory by Dr D. J. Maxwell.

The most important factors in the preparation of HRP-stained neurones for ultrastructural analysis are: (i) the preservation of ultrastructural detail by means of satisfactory fixation of the tissue; (ii) the identification of labelled profiles at the light microscope level and their subsequent examination with the electron microscope; and (iii) the cutting and collection of serial ultra-thin sections, so that the labelled profiles may be examined through their extents and, if desirable, three dimensional reconstructions made at this level of analysis.

The first requirement, high quality fixation to preserve ultrastructural detail, has been discussed in Section II of this Chapter. One point is, perhaps, worth repeating, and that is that the high aldehyde concentrations (3%, especialiy of glutaraldehyde) used for optimal ultrastructural definition lessen the intensity and extent of HRP labelling achieved. This is not usually crucial as long as the injection sites and likely areas containing stained profiles can be accurately located.

The point at which preparation of material for the EM departs from the LM procedure is immediately after the thick sections have undergone the HRP histochemical reaction, i.e. after step 5 in either the DAB or PC-PPD methods given in Section III.C above. In order to avoid having to process an inordinate amount of material through the time-consuming and complicated EM preparation method, it is advisable to select only those sections containing probable HRP-stained material.

The freshly reacted sections (usually 50–60 μm thick) are placed on glass microscope slides under cacodylate buffer solution and viewed with the light microscope. Constant reference to the experimental protocol, photographs of the brain or spinal cord and the locations of marker electrode tracks are necessary aids to the positions and identification of stained profiles. Examination of thick, uncleared wet sections in this way is not easy. Although stained nerve cell bodies can be seen with little effort, even at relatively low magnifications (e.g. × 20–40), fine dendritic branches and especially synaptic boutons on axonal arborizations are difficult to see even at high magnification (× 400). It is useful to view the sections from both sides, as stained profiles near to one surface may be very difficult to see from the other. It is worth bearing in mind that a well-stained neurone (or axon) may have useful and interesting profiles up to 1 cm away from the point of injection, so it is worth scanning a large proportion, or even all, of the thick sections. The sections selected for EM are collected in a numbered sequence in 0.15 M sodium cacodylate buffer (pH 7.6) at room temperature. The remaining sections should be mounted on glass microscope slides, leaving appropriate gaps where sections have been taken for EM, and treated as described earlier (Section III.C.2. steps 6 and 7) for light microscopy.

The sections for ultrastructural analysis are transferred to individual glass vials containing 0.15 M sodium cacodylate buffer and post-fixed for 30 min in a 1% osmium tetroxide solution in cacodylate buffer. They are then washed in six changes of the same buffer and dehydrated in an ascending series of acetone solutions (3 × 5 min in 30%; 10 min in 50%; 10 min in 70%; 2 × 15 min in 100%). The sections are then embedded in Araldite between acetate foils, according to the method of Holländer (1970) (see also, Rastad et al., 1977). The sections are placed on 250 μm thick acetate foils, surrounded by small amounts of unpolymerized Araldite and covered by a 100 μm thick acetate foil. Usually, we mount two to six sections on each pair

68

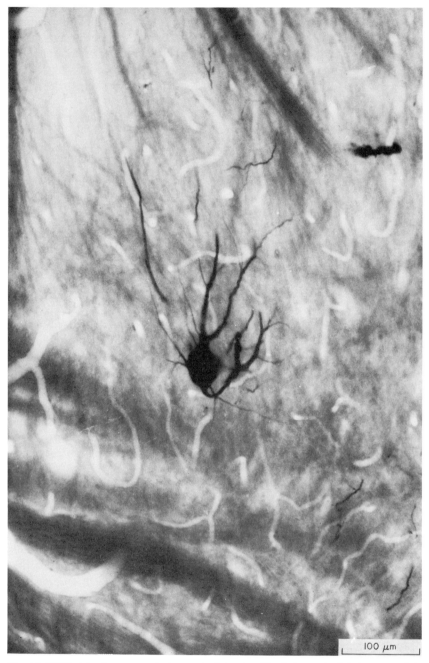

Fig. 16. HRP-stained neurone (spinocervical tract cell) photographed between acetate foils. The section is 50 μm thick and was processed with the PC-PPD method.

of foils, which measure about 7×5 cm. The foils are baked at 60°C for 48h to polymerize the Araldite.

After two days it is once again possible to examine the sections under the light microscope. At this stage, it is essential to make low and high power camera lucida drawings of the stained profiles, including details of the locations of other landmarks, such as blood vessels, bundles of myelinated fibres, neighbouring unstained cell bodies, possible cytoarchitectonic boundaries or the border between grey and white matter, etc., and also to take photomicrographs of the sections (Fig. 16). These drawings and photographs will all be used subsequently, to orientate the electron-micrographs and to allow correlation between light and electron microscopy.

When these LM observations have been completed, the areas of interest selected for EM study are re-embedded. The top acetate foil is stripped off, leaving one surface of the section exposed. A blank Araldite block is now positioned over the part of the section containing stained profiles, using a stereomicroscope for visual control, and attached to the section by a drop of unpolymerized Araldite, maintained in position by applying weights and the whole is transferred to an oven at 60°C to polymerize the Araldite drop. Once polymerization is complete (after two days) the tissue attached to the block can be pulled free from the bottom acetate foil.

The block containing the stained profiles can now be trimmed under microscopical control. Light is transmitted vertically through the block and the stained profiles easily seen within it. This ability to view the block face is a considerable advantage, for example during ultra-thin sectioning the block face may be viewed from time to time, thus obviating the need to cut thicker reference sections from the block.

Ultra-thin sectioning can now be carried out using standard techniques. If possible, serial ultra-thin sections should be cut. The sections are collected on formvar coated grids (one-hole or slot grids being especially useful) and contrasted with uranyl acetate and lead citrate (Reynolds, 1963). The sections are stained with 2% uranyl acetate in 50% ethyl alcohol for 10 min and then in lead citrate for 10 min.

7
Interpretation and Applications

I. INTERPRETATION OF HRP STAINED MATERIAL

A. Quality of Material

1. General

We are often asked what is the rate of our success, that is, what percentage of injections result in stained neurones? It is not an easy question to answer. We certainly have runs of experiments where, for one reason or another (usually to do with problems of microelectrode function, such as poor current passing capability or poor intracellular penetration capability), our yield is disappointingly low. On the other hand, many experiements yield 100% stained neurones. Success is also dependent on the type of neurone or axon that is the target of the experiment.

A number of general points concerning how to obtain high quality histological material have been presented in previous chapters. Some of these will be repeated here together with others that have not been covered.

(*a*) *Perfusion* The necessity for a good perfusion, to clear blood vessels, has been stressed in Chapter 6. The presence of endogenous peroxidase in blood cells can be a great nuisance and may obscure the details of injected neurone (Fig. 17). Endogenous peroxidases in neuronal elements (Bunt *et al.*, 1974) are less of a problem in injected material than in retrograde or anterograde transport studies, since the injected neurones are so densely stained.

(*b*) *Damage occurring during the electrophysiological experiments* Several types of damage may occur that affect the quality of the final histological material. Surface damage where the pia-arachnoid has been removed to allow microelectrode penetration may lead to local extravasation of blood. Indeed repeated microelectrode tracking through a small pial hole may lead to the same problem, as may inadvertent puncture of a surface vessel by the electrode

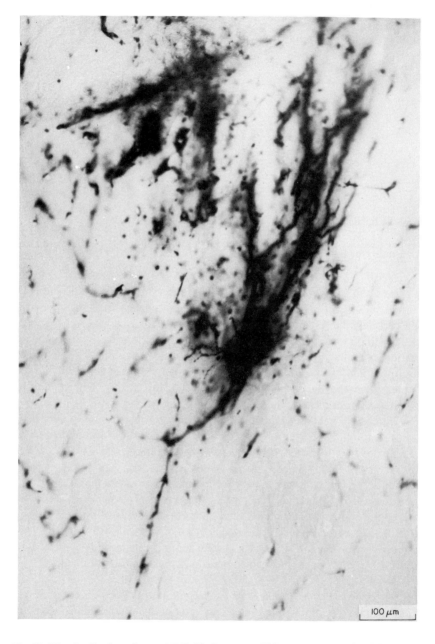

Fig. 17. Blood cells obscuring an HRP-filled neurone. This was a preparation in which the perfusion was of poor quality and, in addition, there was a small haemorrhage near the stained cell. (Cat spinal cord, 100 μm thick transverse section, PC-PPD reaction).

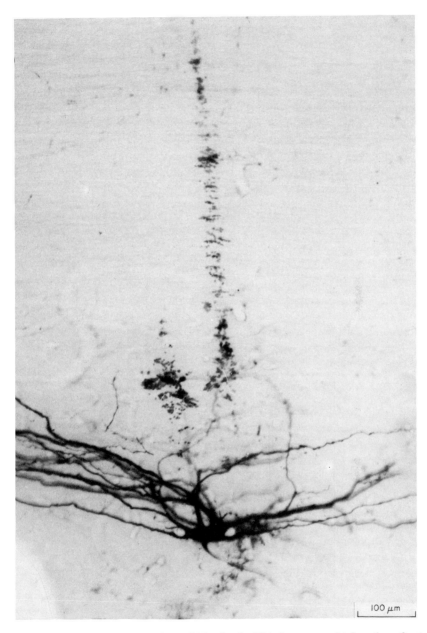

Fig. 18. Electrode tracks containing red blood cells. This is a parasagittal section of cat spinal cord (100 μm thick section, PC-PPD reaction), with a stained dorsal spinocerebellar tract neurone at the bottom of the figure. Above the neurone, parts of two micro-electrode tracks can clearly be seen.

when the damage is usually more severe. These surface blood cell accumulations should not be a nuisance if the target neurones are located at some depth from the surface. We tend to avoid making injections very close to the surface whenever possible, not only to avoid the problems mentioned in this section, but also because penetrations (e.g. axonal penetrations) at the surface tend to be less stable than those made slightly deeper, and in the spinal cord, surface injections of axons increase the distance to the terminal boutons within the grey matter. If a microelectrode breaks within a track, you should assume a disaster has taken place from the point of view of subsequent histology. Inevitably, there will be a large spillage of HRP and the only satisfactory recourse is to move well away (several mm) from the position of the track where the accident happened.

(c) *Electrode tracks* The presence of electrode tracks may be useful or may be a nuisance. They are usually visible, due to small accumulations of red blood cells within them, or to very small extracellular deposits of HRP (Fig. 18). They can occur with either very fine or with coarse electrodes and, subjectively, we have the impression that they are more common if very low rates of tracking (e.g. 2 μm steps with pauses between steps) have been used. Electrode tracks may usefully act as markers, especially where detailed protocols have been made during the tracking and if more than one neurone has been injected within a small area. However, the use of blank marker electrodes and photographs of the surface of the brain or spinal cord, as described in Chapter 4, is a safer and cleaner way of providing detailed information on the location of stained neurones. In general, we prefer not to see electrode tracks, other than those produced by the marker electrodes. Sometimes, the sections split at the position of electrode tracks with disastrous results (Fig. 19).

2. Assessment of staining quality
In general, a cell or an axon may be under-stained, well-stained or over-stained. The problem is to decide which of these states has occurred with each neurone. Gross under- or over-staining are easily recognized, but slight under- or over-staining are more difficult to differentiate.

(a) *Under-staining* In the limit, the most under-stained neurone is one that you cannot find in the histological material! This can be more serious than it appears at first sight. If a number of cells have been injected in close proximity to one another, and this may be a necessary condition in the experiment, then the loss of one or more of them can ruin what could have been a very fine set of results. The use of photographs and marker electrodes will help, as mentioned previously.

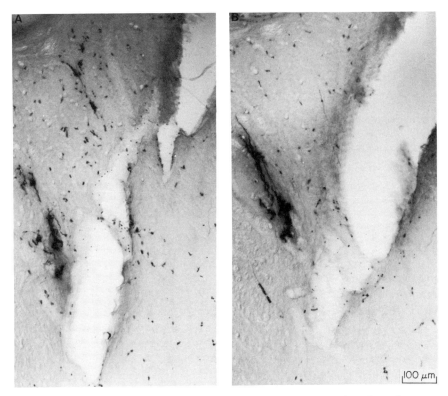

Fig. 19. Splitting of sections along electrode tracks. Two adjacent sections from the same series to show a split along a microelectrode track. To the left of the split, parts of the dendritic tree of an HRP-injected neurone can be seen. The split in the section made reconstruction of this neurone impossible. (Cat spinal cord, 100 μm thick sections.)

Poorly stained neurones are generally easy to identify. The staining is faint and often can be seen to be limited only to the soma and proximal dendrites of the cell (Fig. 20A). Under-stained axons are also faint and HRP reaction product is limited to a short length (a few mm) and any collaterals given off from the axon will be faint also and not stained to the level of terminal boutons (Fig. 20B), although some boutons may be visible on some branches. When under-staining is less obvious than this, it may be useful to compare the material with Golgi pictures of the same, or a similar, set of neurones. As mentioned previously, the Golgi methods do not give as complete a picture of a neurone as a good HRP fill, so any suggestion that a presumed under-stained HRP-injected neurone has fewer dendrites or less well branched dendrites or less extensive dendrites than the Golgi picture will indicate that under-staining is probable. Obviously, any dendrite or axon branch that shows fading of reaction product rather than a clearly defined end should lead to the suspicion that the neurone or axon is under-stained. The use of intensification procedures (see Chapter 6) helps to overcome some of the problems of under-staining, and it is recommended that an intensification is carried out on tissue where any suspicion of possible under-staining exists (for example after only a small amount of injection current has been passed). However, if the same tissue block or sections contain material that should be well-stained then intensification may lead to problems of over-staining.

There are uses for material that appears under-stained (or lightly stained) at the light microscope level. Such material may be really excellent for electron microscopical analysis, since intracellular detail will not be obscured

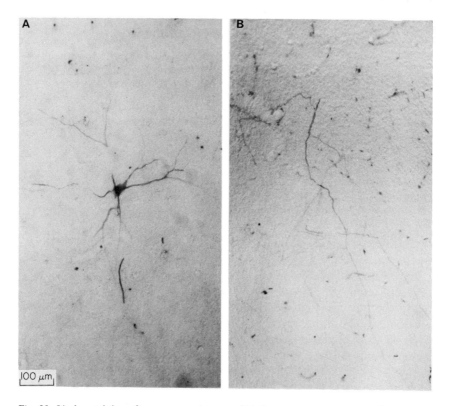

Fig. 20. Under-staining of neurones and axons. (A) shows an α-motoneurone in a 100 μm thick, parasagittal section of cat lumbosacral spinal cord (PC-PPD reaction). Note the light staining of the dendrites which fade as they are traced distally. (B) shows under-staining of a primary afferent fibre collateral in a 100 μm thick transverse section of cat spinal cord. Note the light staining of the collateral branches and that they fade as they pass ventrally (towards the bottom of the figure). No synaptic boutons are visible.

by the HRP reaction product, as it so often is in well-stained material. An example of a slightly stained neurone as seen in the electron microscope is shown in Fig. 21 where post-synaptic thickenings may be seen clearly.

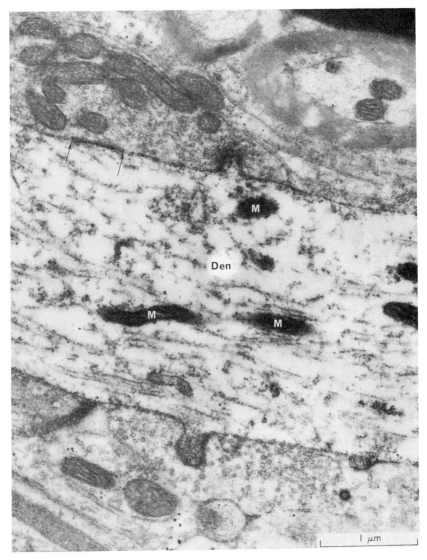

Fig. 21. Electronmicrograph of lightly stained neurone. A lightly stained dendrite (Den) of a dorsal spinocerebellar tract neurone is shown, containing HRP reaction product, which has accumulated more densely in mitochondria (M). Considerable intracellular detail is apparent, including a post-synaptic density (between the arrows). The electronmicrograph was prepared by Dr D. J. Maxwell.

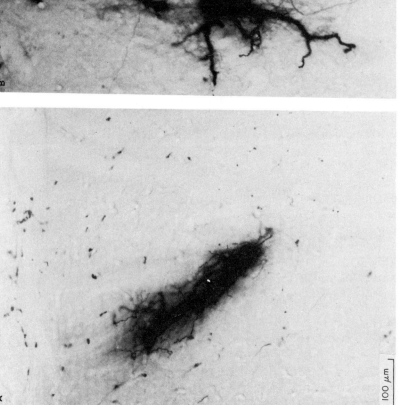

100 μm

Fig. 22. Over-staining of HRP-injected neurones. (A) shows a neurone that has been over-injected (80 nA. min) with some leakage of HRP reaction product around the cell. The dense staining would make this neurone difficult to reconstruct. (B) shows a grossly over-injected neurone that is distorted with the cell body "blown up" by the injection. Both micrographs are from 100 μm thick, transverse sections of cat spinal cord reacted with the PC-PPD method.

(b) *Over-staining* Over-staining produces two main problems: distortion of the neurone and leakage of HRP around the neurone. In our experience, over-staining of injected axons does not occur, the main problem with axons being under-staining or, at least, not as much staining as one would like to see. Over-staining of neurones is difficult to predict until considerable experience has been gained of the neuronal type under study. In general, the use of coarse electrodes and large amplitude injection currents (greater than 20 nA) and the injection of more than 100 nA.min of current predispose to over-staining, especially of small to moderate sized neurones. Large neurones, such as α-motoneurones, can often cope with coarse electrodes, large current amplitudes and large total currents.

Grossly over-stained and distorted neurones are generally easy to recognize (Fig. 22). Less gross distortion provides more difficulties and comparison with high quality Golgi material may be helpful. A scalloped appearance of the neuronal soma probably indicates distortion. However, distortion *per se* is not necessarily due to over-staining. It is advisable, especially if a number of cells appear to be distorted, to consider other possible causes such as the osmolarity of the perfusion fluids and histological reagents or the effects of freezing unprotected tissue. Most of the distortion due to over-staining (over-injection) presents itself as disruption of the neurone with ballooning of the soma and/or dendrites (Fig. 22B).

In addition to cell disruption, over-staining is usually indicated by leakage of the HRP, leading to the appearance of HRP reaction product around the cell (Fig. 22). This leakage may be present by itself, with no obvious cell disruption. However, in our experience, the two signs of over-injection are nearly always present together and make interpretation, especially cell reconstruction, difficult if not impossible. Any HRP that leaks around the cell is taken up by neighbouring cells (neurones and glia) (Fig. 23). In extreme cases it is even impossible to be sure which neurone has been injected. More often, however, the injected neurone is the most densely stained one of the group (Fig. 36). As mentioned above, using intensification procedures on tissue such as this will exacerbate the situation.

(c) *Good staining* The aim should be to produce staining that is uniform and of a high contrast between the stained cell and the background. For dendritic trees (injections into cell bodies) uniform density of stain should certainly be the objective (Fig. 24). Usually the staining intensity is somewhat greater in the soma and proximal dendrites, but any fading to distal dendrites should be slight and dendritic branches down to diameters of 1 μm or less should be easily visualized.

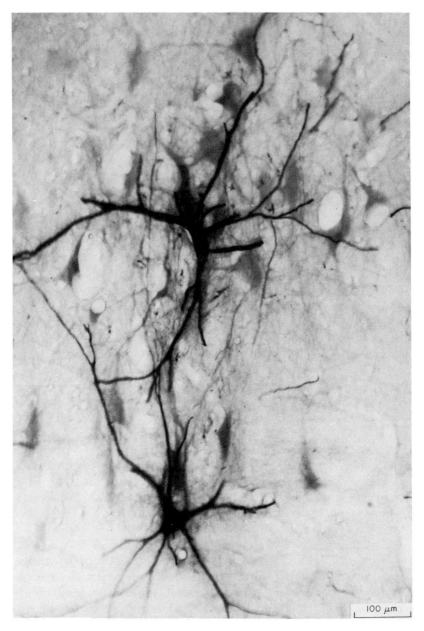

Fig. 23. Leakage of HRP around stained neurones. Two α-motoneurones in the cat's lumbosacral spinal cord (100 μm thick, parasagittal sections, PC-PPD reaction) stained by intracellular injection. They are surrounded by many neurones, and some axons, that have taken up the HRP in amounts sufficient to prevent reconstruction of the injected neurones.

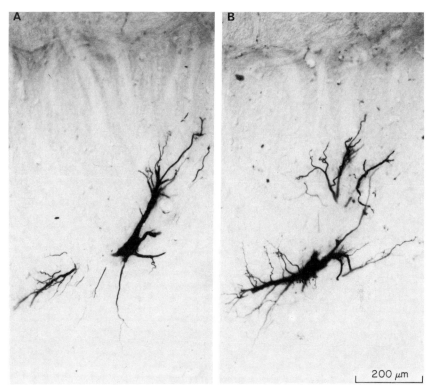

Fig. 24. Well-stained neurones. (A and B) show two adjacent transverse sections (100 μm thick, PC-PPD reaction) of a neurone in lamina III of the cat's spinal cord. Note the even staining and the "clean" edge to the dendrites.

Neurones known to have dendritic spines or other appendages should exhibit them in HRP material (Fig. 25). Indeed, in our experience, HRP-stained neurones are more likely to show dendritic spines than are Golgi preparations of the same types of cells. The cell's axon should also be identifiable (except where it ramifies exclusively within the dendritic tree—see below) and stained for several mm (Fig. 26). When axons are injected, a complete fill of all the terminal arborizations is highly unlikely. However, these terminal arbors that are filled should show a uniform density of staining with no faint branches and the fine branches within the arbor should display synaptic boutons (Fig. 27). Fine terminal branches should, as far as can be ascertained from serial sections, terminate in a bouton and should not fade away.

The thick myelinated parts of axons, however, very rarely show even staining, especially when the PC-PPD reaction is used (Fig. 26B). The DAB method produces more evenly stained axons (Fig. 26A). The myelin appears to obstruct penetration by the histochemical reagents and myelinated axons

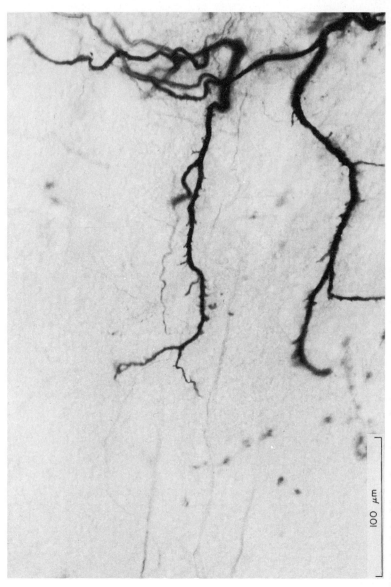

Fig. 25. Well-stained dendrites of a spinocervical tract neurone (100 μm thick sections, PC-PPD reaction). Note the numerous dendritic spines. The fine profiles running (mainly) from left to right across the micrograph are branches of a hair follicle afferent collateral that was also stained in this experiment.

100 μm

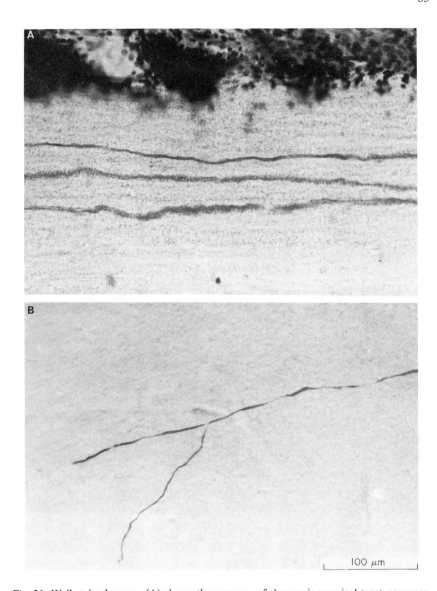

Fig. 26. Well-stained axons. (A) shows three axons, of three spinocervical tract neurones in cat, running in the dorsolateral funiculus of the spinal cord. The photomicrograph is of a 50 μm thick parasagittal section reacted with the DAB method after formaldehyde fixation. Only the upper axon is in focus. Note the absence of obvious nodes and the relatively even staining. (B) shows a collateral from a Ia muscle spindle afferent fibre (cat) running through the spinal cord grey matter and giving off a branch. This is a 100 μm thick section reacted by the PC-PPD method after formaldehyde-glutaraldehyde fixation. Note the variation in thickness of the axon with nodes clearly visible. The branch is given off at a node.

84

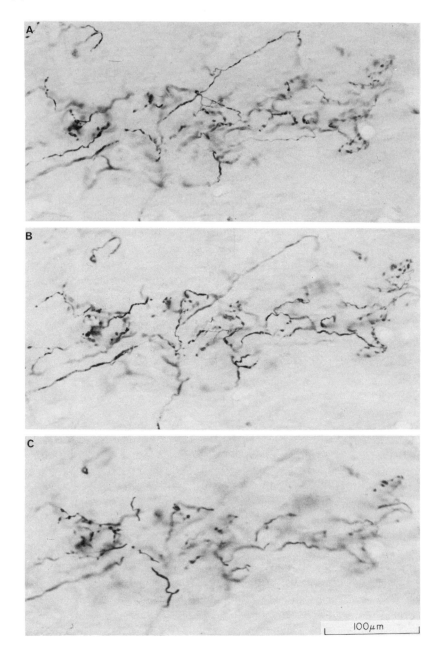

Fig. 27. Well-stained axon collaterals carrying synaptic boutons. (A, B and C) show a series of photomicrographs at different levels of focus through the same 100 μm thick section (cat spinal cord, PC-PPD reaction).

are often only densely stained at the nodes, with faint or even absent staining between them. The thickness of the section also plays a part here; 100 μm thick sections show greater staining irregularities in myelinated axons than 50 μm sections. Also, the greater the aldehyde concentrations in the fixative then the greater the staining irregularities. Anything that tends to inhibit the histochemical reaction will exacerbate this problem.

3. Artefacts

Apart from obviously spurious results due, for example, to HRP leakage from an electrode tip or to leakage from an over-injected or damaged neurone, there are two possible sets of artefacts produced by the HRP method. They may not be artefacts at all, but at present it is difficult to be certain. These two sets of phenomena are "beads" or varicosities on dendrites and the occurrence of more than one cell stained with no other signs of HRP leakage.

(a) "Beads" and varicosities A common feature of HRP filled neurones is the appearance of strings of bead-like enlargements along the terminal parts of dendrites (Fig. 28A). They are especially prominent in α-motoneurones and dorsal spinocerebellar tract neurones (Cullheim and Kellerth, 1978; Burke et al., 1979; Brown and Fyffe, 1981; Rose, 1981; Rose and Richmond, 1981; Houchin et al., 1983). Similar beading has been seen in Golgi preparations (see for example Rose and Richmond, 1981). In the electron microscope, these "beads" are usually seen to contain mitochondria and are often the site of obvious intracellular disruption (Fig. 28B). We suspect that the beads are in many cases, artefactual.

(b) Multiple staining It is essential, in order to make meaningful correlations between structure and function, that the neurone visualized by HRP injection is the one from which electrophysiological recordings were made. We have already stressed the need to take precautions to ensure that this is so, as far as possible (Chapter 4). But what if two (or more) cells are labelled where only one was thought to have been injected and there are no obvious signs of HRP leakage around one cell? This does occasionally happen in our experience (Fig. 29), but it is our opinion that the cause is usually due to inadvertent injection of two neurones. Thus, an injection may be started, the cell may be lost, the electrode manipulated and apparently repenetrates the same cell as far as can be determined (although a complete re-run of identification procedures may not be carried out) and the injection continued. In this situation, a double labelling of a pair of cells should immediately lead to the suspicion that two cells have been injected. If continuous intracellular monitoring during current injection is not performed any multiple labelling should be suspect. In our experience, we have only very rarely seen double

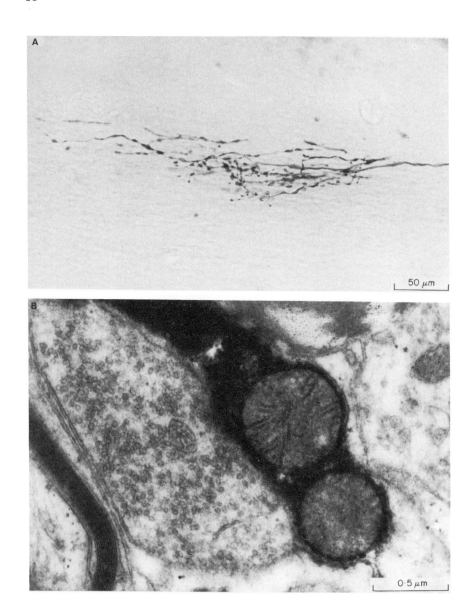

Fig. 28. Beading of distal dendrites. (A) is a light micrograph to show the typical appearance "beads" (dorsal spinocerebellar tract neurone, parasagittal section, 100 μm thick, PC-PPD reaction). (B) is an electronmicrograph from another dorsal spinocerebellar tract neurone, similarly reacted, and shows the swollen mitochondria characteristically contained within "beads". The electronmicrograph was prepared by Dr D. J. Maxwell.

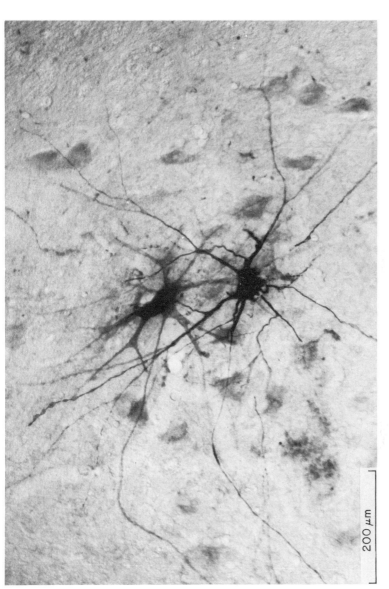

200 μm

Fig. 29. Staining of two neurones after injecting one. In this example two α-motoneurones are stained, one fainter than the other, even though electrophysiological controls indicated that only one had been penetrated and injected. In such a situation (which is very rare in spinal cord), it is impossible to know which neurone was injected or if the microelectrode passed from one to the other. (Cat lumbosacral spinal cord, 100 μm thick, parasagittal section, PC-PPD reaction.)

labelling in the spinal cord when we thought that a single cell had been injected. However, it is conceivable that if there are gap junctions between cells then HRP could be passed from one to another. Only ultrastructural examination can answer this problem. Other tracers do seem to pass more readily from cell to cell, for example, ^{3}H-glycine (Globus *et al.*, 1968; Schubert, 1974; Kreutzberg *et al.*, 1975) and Procion Yellow (Zieglgänsberger and Reiter, 1974).

There is one form of multiple labelling, however, that cannot be due to double injections, and which also seems unlikely to be due to simple leakage of HRP from an over-injected neurone. This can be called labelling at a distance, and has been noted by Hongo *et al.* (1981) and by ourselves after tissue has been intensified with the cobalt method. An example is shown in Fig. 30, where a primary afferent fibre was injected near the dorsal root entrance zone of the spinal cord. The fibre innervated a slowly adapting Type II mechanoreceptor and its collaterals distributed to the medial grey matter of the dorsal horn. Many boutons may be seen carried on the terminal arborization and some are clustered around a neurone that has also taken up HRP. The injection was made several mm away from the neurone that has taken up HRP and there are no signs indicative of extracellular HRP leakage. The mechanism of such take-up by post-synaptic elements is unclear, for example, why did this particular neurone take up the HRP when others apparently did not (or at least did not in enough quantity to reveal reaction product with PC-PPD and cobalt intensification methods)? Interneuronal transfer may be the means by which such a phenomenon occurs, but its mechanism is at present unknown. Whatever the mechanism, however, this phenomenon of labelling at a distance may well be useful in the study of monosynaptic connexions.

4. Injection sites

The site of injection of a neurone may often be determined. Indeed in some circumstances it is most useful, and occasionally necessary, to be able to recognize injection sites. There are, of course, electrophysiological signs that indicate the location of an electrode tip and when combined with knowledge of the cellular organization in the area under study these signs will give a secure foundation for such location. Thus, dendritic impalements in pyramidal cells of the cerebral or cerebellar cortices may be recognized (e.g. Ekerot and Oscarsson, 1981). Axonal impalements near to the cell body may also be recognized, by the presence of small amplitude synaptic potentials and step-wise changes in amplitude and shape of action potentials evoked at decreasing series of short intervals (Gogan *et al.*, 1983), for example. Identification of the injection site in the histological material will provide conclusive evidence for these more tentative localizations made during the electrophysiological experiments.

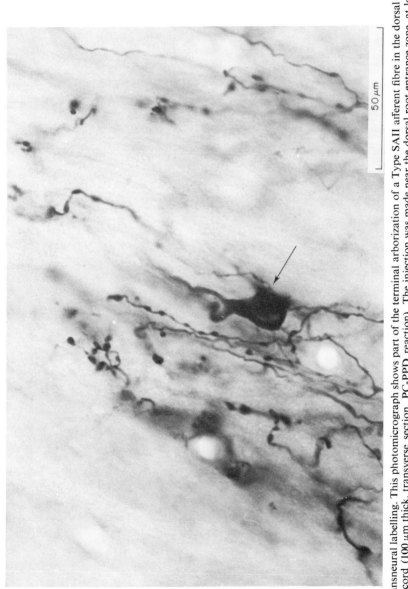

50 μm

Fig. 30. Transneural labelling. This photomicrograph shows part of the terminal arborization of a Type SAII afferent fibre in the dorsal horn of the cat's spinal cord (100 μm thick, transverse section, PC-PPD reaction). The injection was made near the dorsal root entrance zone, at least 1.5 mm away from the terminals and yet a neurone (arrow) has been stained. The sections were taken through the cobalt intensification procedures.

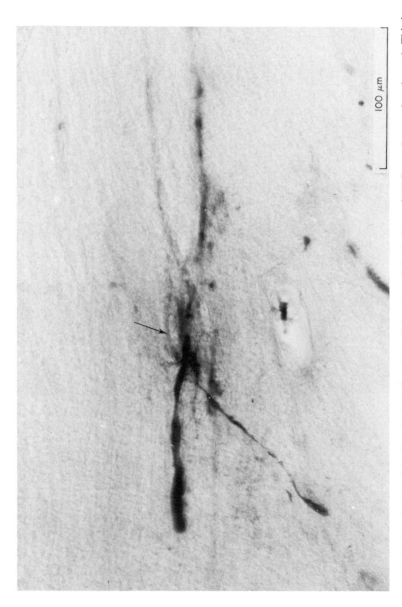

100 μm

Fig. 31. Axonal injection site. The injection site is indicated by slight extracellular leakage of **HRP** reaction product (arrow). This is a parasagittal section of cat spinal cord (100 μm thick) showing a Ia muscle afferent fibre running from right to left within the dorsal columns. A collateral is given off close to the injection site.

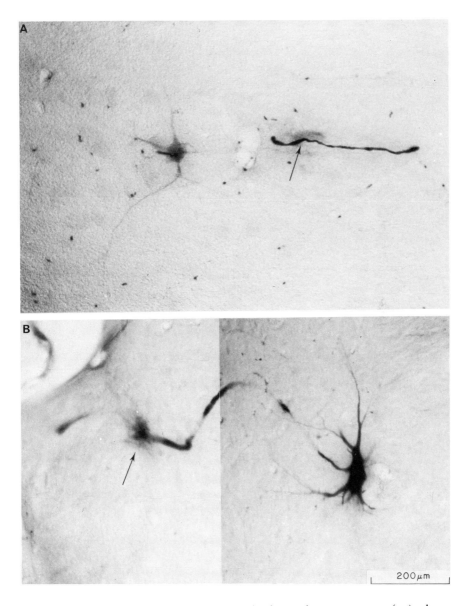

Fig. 32. Intra-axonal injections. (A) shows a lumbosacral α-motoneurone (cat) whose axon was injected (at the arrow). The axon is densely stained, but the soma is very lightly stained and only the most proximal parts of the dendritic tree appear. (Parasagittal section, 100 μm thick, dorsal to the left). (B and C) show two adjacent sections (100 μm thick, transverse) of cat spinal cord with a probable ventral spinocerebellar tract neurone. The injection was made into the axon (arrow in B) and in this case the cell is reasonably well-stained. PC-PPD reaction for both neurones.

Injection sites may be identified histologically where the electrode penetration or current passage cause some distortion or HRP leakage. At injection sites of axons there is usually some swelling and slight extra-axonal deposition of HRP reaction product (Fig. 31). This is a good marker and seldom causes any problems by staining other fibres (although the use of intensification procedures can increase the staining intensity of axons that have taken up leaked HRP). Dendritic injection sites are similarly indicated by swelling and slight HRP leakage.

When axons are injected close to a neurone's soma there is usually clear histological evidence for this (Fig. 32). The soma and dendrites are much more faintly stained than the axon and its collaterals.

5. Distinguishing between axonal and dendritic profiles

Where the neurone has a thick (myelinated) axon there is usually little problem in recognizing it and tracing its collaterals from their point of origin. The initial few hundred μm of such an axon from its origin is often poorly stained, however, and it may be difficult or impossible to recognize it (Fig. 33). Thick myelinated axons are often stained irregularly, the myelin apparently acting as a barrier to the histochemical reagents, which only seem to gain access at nodes (see above). Since axon collaterals are given off at nodes they are recognizable. Dendrites should be capable of reconstruction back to the cell's soma.

There are a few general guide lines for the recognition of axon collaterals vs. dendrites. Axon collaterals and their preterminal branches (Fig. 34) usually have a smooth profile, are uniformly stained and show no tapering. Where they are (thinly) myelinated there may be some slight variation in staining intensity along their length (Fig. 26B), with the more intense staining being at the nodes (where branching takes place). Terminal axonal branches will usually carry boutons *en passant*, although occasionally only boutons *terminaux* will be found. Dendrites often carry spines and other protuberances (Fig. 25) and branch at web-like divisions (Fig. 35). The angle between branches is often fairly acute (20–45°), whereas axonal branching is more often at larger angles (60–90°) (see Figs. 26B, 34, 36). The diameter of daughter branches is not often a very useful guide. Although thick axons will often give rise to thin branches, this is not necessarily so. Preterminal axons often divide into fairly equal daughter branches and dendrites seldom divide equally.

Most difficulties in distinguishing between axonal and dendritic profiles occur when the axon ramifies within or close to the dendritic tree of the parent neurone. Here it can be impossible to make the distinction, especially where terminal dendrites have a beaded appearance. Ultimately, in this sort of situation, only electron microscopical analysis will provide an answer, and

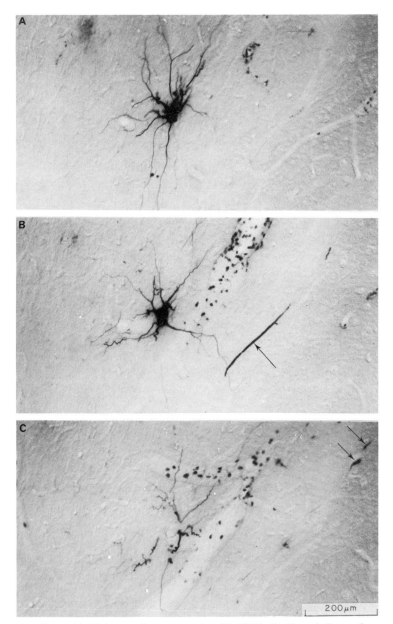

Fig. 33. Failure of initial part of axon to stain with HRP. (A, B, and C) are three serial sections from the same neurone (100 μm thick sections, cat spinal cord). The axon appears well-stained in (B) (arrow) and its distal continuation may be seen in (C) (arrows). The initial part, however, cannot be seen in (A or B) (nor in any of the other sections from the series) and has not been stained.

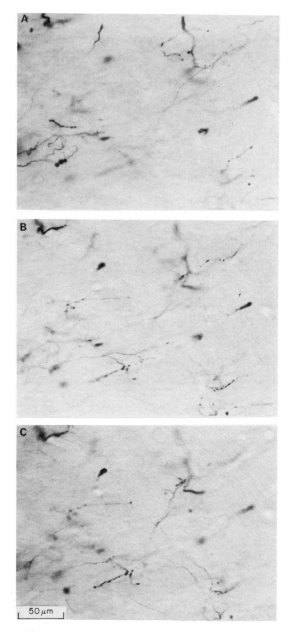

Fig. 34. Axonal profiles. (A, B and C) are photomicrographs of the same 100 μm thick section taken at different field planes. Note the fine, bouton-carrying, terminal axonal profiles, the smooth surface to the thicker branches and, in general, branching occurring at angles approaching 90°.

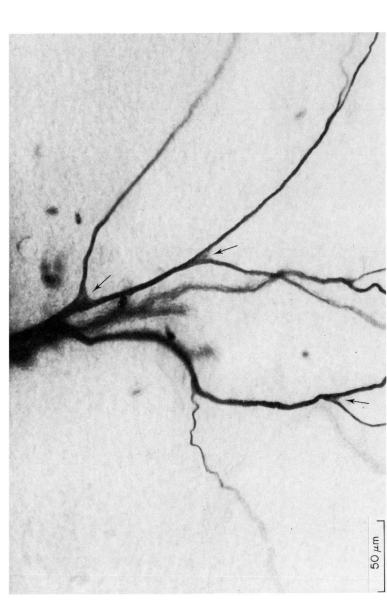

Fig. 35. Dendritic profiles. The micrograph shows dendrites of a cat α-motoneurone. The dendrites do not carry spines (cf. Fig. 27). Note the angles at which branching occurs (less than 90°), usually about 60°), with generally roughly equal diameters for the daughter branches which are thinner than the parent branches. Note also the web-like infilling of some of the branch points (arrows).

50 μm

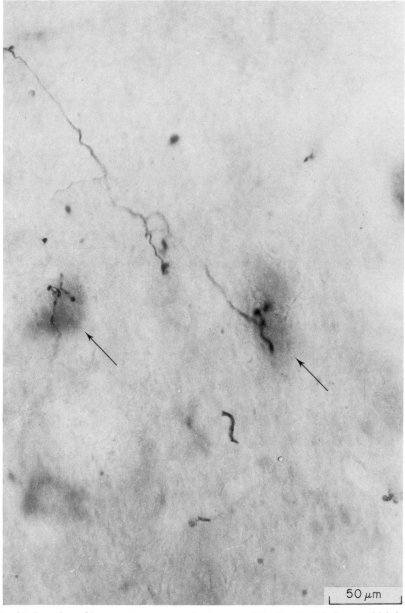

Fig. 36. Location of boutons in counterstained nerve cell bodies. In this figure, which is a 100 μm thick transverse section of cat lumbosacral spinal cord, there is an HRP-stained branch of a Ia muscle afferent fibre collateral running diagonally across the micrograph (PC-PPD reaction). The section was counter-stained with methyl green (Nissl stain) and boutons from the Ia afferent can be seen overlying two α-motoneuronal cell bodies.

then only when the staining is sufficiently diffuse to allow intracellular detail to be clearly seen.

6. Ultrastructural interpretation

There are no special problems of interpretation at the electron microscopical level of HRP-stained material, other than the obscuring of intracellular detail by the reaction product, as long as the quality of the fixation is high. The reader is referred to standard references on the ultrastructure of nervous tissue (e.g. Peters et al., 1976).

II. APPLICATIONS

Apart from the obvious use of the intracellular HRP method to provide light and electron microscopical material from neurones that have been recorded electrophysiologically, the method may be used in several ways and combined with a number of other techniques. In this way its power, which is already considerable, can be made even greater. When combined with counterstaining (Nissl or silver stains) the location of probable contents between neurones may be determined (Fig. 36). In the following sections we will mention some additional possibilities.

A. Simple Applications

By simple applications we mean the use of the intracellular HRP technique by itself without additional procedures. We do not mean that the techniques are not difficult. In fact, the following examples were all extremely difficult technically.

1. Small axons and neurones

The use of very high impedance microelectrodes combined with a high voltage electrometer allowed Light and Perl (1979), Light et al. (1979), Mense et al. (1981) and Light et al. (1981) to inject fine myelinated primary afferent fibres and substantia gelatinosa neurones in the cat's spinal cord in the very short time during which the microelectrode was intracellular. These were very considerable achievements. It should be noted, however, that others have succeeded in staining small neurones in the substantia gelatinosa with more conventional set-ups, although usually with rather high impedance electrodes (Bennett et al., 1979, 1980; Molony et al., 1981), and also small neurones (basket cells) in the visual cortex (e.g. Martin and Whitteridge, 1982).

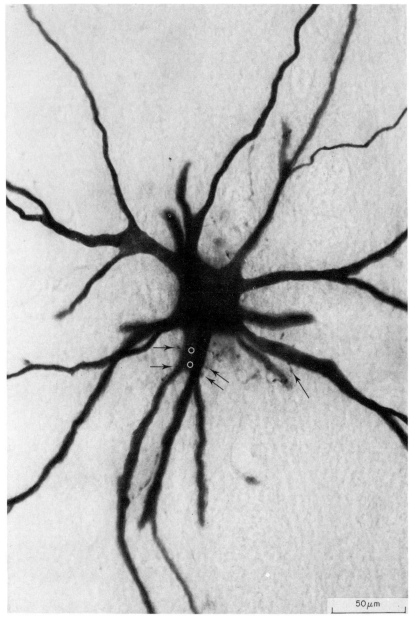

Fig. 37. Double staining to show synaptic contacts. The figure is an oil-immersion photo-micrograph showing five contacts (arrows) between a Ia muscle afferent fibre and an α-motoneurone in the cat's lumbosacral spinal cord (100 μm thick section, PC-PPD reaction). From Brown and Fyffe, 1981.

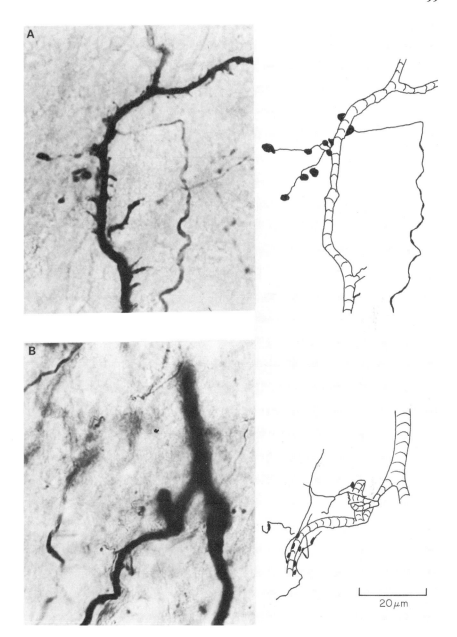

Fig. 38. Double staining to show synaptic contacts. Photomicrographs, on the left, and camera lucida drawings, on the right, of synaptic contacts between hair follicle afferent fibres and spinocervical tract dendrites. Some of the contacts are indicated by arrows (100 μm thick sections, PC-PPD reaction, cat spinal cord). From Brown and Noble, 1982.

2. Distant terminals of ascending and descending pathways
In these applications, long axons have been injected near their site of termination. Thus, corticospinal axons have been injected in the cervical (Futami *et al.*, 1979) and lumbar (Brown *et al.*, 1982) spinal cords and their terminal arborizations studied at light and electronmicroscopical levels. Similarly, descending raphe-spinal axons have also been studied after injections at the lumbar level (Light, 1983). Thalamo-cortical afferent fibre arborizations have also been studied in the cerebral cortex (Deschênes and Landry, 1980).

3. Pairs of neuronal elements
By injecting pairs of neuronal elements known to be connected, the numbers and locations of synaptic boutons from one member onto the other may be determined. This has been done for the Ia afferent fibre-α-motoneurone connexion in the feline spinal cord (Burke *et al.*, 1979; Brown and Fyffe, 1981; Fig. 37), for the connexion between Ia afferent fibres and neurones of the dorsal spinocerebellar tract in Clarke's column (Tracey and Walmsley, 1982), and for hair follicle afferent fibres and spinocervical tract neurone (Brown and Noble, 1982, Fig. 38). If the two neuronal elements can be recorded simultaneously and then injected (Redman and Walmsley, 1983), not only may the numbers and locations of synaptic boutons be determined but also the electrophysiological parameters of the synaptic potentials evoked in one by the other may be measured, with important implications for neuronal modelling studies. In experiments where pairs of elements are injected (either sequentially or simultaneously), the ultimate test for synaptic connectivity between them is a demonstration at the electron microscope level.

B. Combination with Other Techniques

1. Combination with injection of other intracellular stains
This is really a variant on Section IIA.3 above. In our original report of the HRP method (Snow *et al.*, 1976) we included results from an experiment where one neurone had been injected with HRP and another with Procion Yellow. It was possible to observe HRP-filled boutons in close apposition to the soma and dendrites of the Procion Yellow-filled neurone (see Fig. 39). The only advantage of using two markers in this way, as opposed to a double injection of HRP, is where contacts are most likely to be on the soma and proximal dendrites of a neurone. In this case, they are more likely to be seen if the post-synaptic neurone is injected with a dye that provides good contrast with the HRP reaction product.

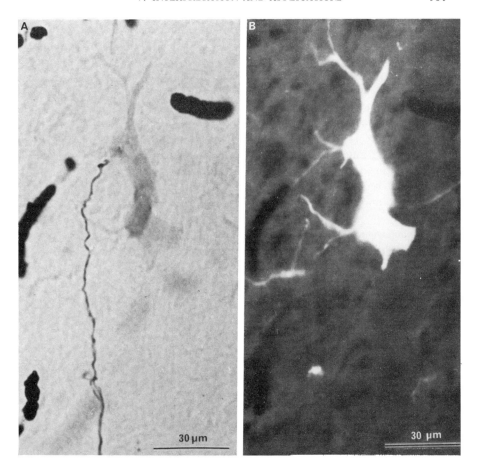

Fig. 39. Double staining with Procion Yellow and HRP. This pair of photomicrographs shows the same tissue, (A) photographed with tungsten illumination and (B) with fluorescence illumination. Procion Yellow had been injected into the neurone whose cell body appears in the upper half of each micrograph and HRP into another neurone, whose axon gave off collaterals that made contact with the Procion Yellow injected cell (arrows.) Modified from Snow *et al.*, 1976.

2. Combination with retrograde and anterograde tracers

The combination of HRP injection of neurones (or axons) with staining of target neurones by retrograde or anterograde tracers of various kinds promises to be a powerful technique. Thus, the use of the retrograde HRP technique to stain neurones with long axons while in the same experiment injecting neurones that make synaptic contact with them will provide important anatomical information on the connexions. This has been done by Tracey and Walmsley (1982), who marked dorsal spinocerebellar tract neurones by intracerebellar injection of HRP and then made intra-axonal injections into Group Ia muscle afferent fibres near Clarke's column. We (B. A. Bannatyne, D. J. Maxwell, R. Noble and A. G. Brown, unpublished observations) have performed similar experiments, in which post-synaptic dorsal column neurones have been labelled retrogradely and primary afferent fibres have been injected intra-axonally (Fig. 40).

There are other combinations that appear promising. For example, Somogyi *et al.* (1979) have combined Golgi staining with retrograde HRP labelling and anterograde degeneration techniques. In our laboratory, we have combined anterograde degeneration with intracellular HRP injection (Fig. 41). Combination of intracellular HRP injection with autoradiography would also appear to be straightforward.

3. Combination with transmitter identification
A major thrust in many laboratories at the moment is the development of techniques for combining intracellular HRP injection with transmitter histochemistry of the tissue. At the time of writing, we do not know of any

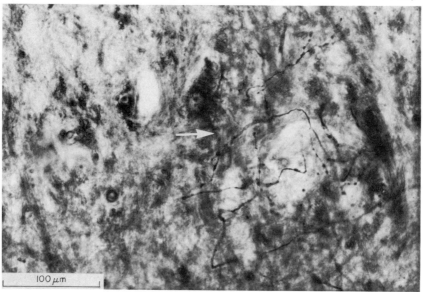

Fig. 40. Combination of retrograde HRP transport with HRP injection. The photomicrograph shows a retrogradely labelled neurone of the post-synaptic dorsal column system and a terminal arborization from a primary cutaneous afferent fibre with boutons contacting the neurone (arrow). (Cat spinal cord, 100 μm thick parasagittal section, PC-PPD reaction and cobalt intensification.) (From unpublished results of B. A. Bannatyne, A. G. Brown, D. J. Maxwell and R. Noble.)

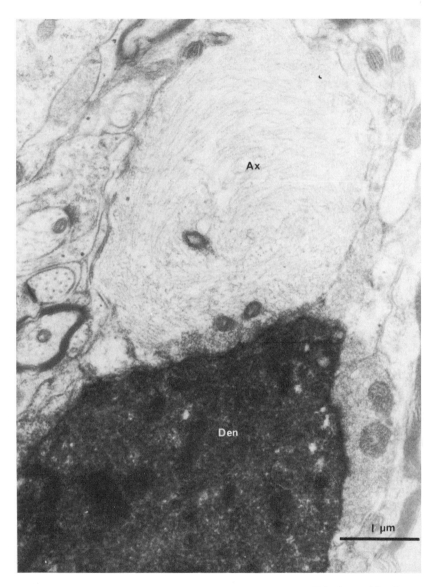

Fig. 41. Combination of anterograde degeneration with HRP injection. This electron micrograph shows a dendrite (Den) of a spinocervical tract neurone, which had been injected with HRP, receiving degenerating terminals of primary afferent fibres (Ax). (From unpublished results of A. G. Brown, R. E. W. Fyffe and D. J. Maxwell, micrograph prepared by D. J. Maxwell.)

reports of the successful use of such a combination. However, Bowker *et al.* (1981) and Yezierski *et al.* (1982) have successfully combined retrograde HRP studies with serotonin immunocytochemistry. It would appear to be only a matter of time, and possibly as short a time as it takes this book to be published, before such combinations become generally available. When that happens, a new era in combined neuroanatomical, neurophysiological and neuropharmacological studies will be upon us.

References

Adams, J. C. (1977). Technical consideration on the use of horseradish peroxidase as a neuronal marker. *Neuroscience* **2**, 141–145.

Adrian, E. D. and Moruzzi, G. (1939). Impulses in the pyramidal tract. *J. Physiol.* **97**, 153–199.

Andersson, S. A. and Källström, Y. (1971). A closed chamber for microelectrode recording from the brain. *Acta physiol. scand.* **82**, 3–4A.

Barrett, J. N. and Crill, W. E. (1974a). Specific membrane properties of cat motoneurones. *J. Physiol.* **239**, 302–324.

Barrett, J. N. and Crill, W. E. (1974b). Influence of dendritic location and membrane properties on the effectiveness of synapses on cat motoneurones. *J. Physiol.* **239**, 325–345.

Barrett, J. N. and Graubard, K. (1970). Fluorescent staining of cat motoneurons *in vivo* with bevelled micropipettes. *Brain Res.* **18**, 565–568.

Barrett, J. N. and Whitlock, D. G. (1973). Technique for beveling glass microelectrodes. *In* "Intracellular Staining in Neurobiology" (S. B. Kater and C. Nicholson, eds), pp. 297–299. Springer Verlag, Berlin.

Barron, D. and Matthews, B. H. C. (1938). The interpretation of potential changes in the spinal cord. *J. Physiol.* **92**, 276–321.

Bennett, G. J., Hayashi, H., Abdelmoumene, M. and Dubner, R. (1979). Physiological properties of stalked cells of the substantia gelatinosa intracellularly stained with horseradish peroxidase. *Brain Res.* **164**, 285–289.

Bennett, G. J., Abdelmoumene, M., Hayashi, H. and Dubner, R. (1980). Physiology and morphology of substantia gelatinosa neurons intracellularly stained with horseradish peroxidase. *J. Comp. Neurol.* **194**, 809–827.

Berthold, C. H., Kellerth, J. O. and Conradi, S. (1979). Electron microscopic studies of serially sectioned cat spinal alpha-motoneurons. I. Effects of microelectrode impalement and intracellular staining with the fluorescent dye "Procion Yellow". *J. Comp. Neurol.* **184**, 709–740.

Bowker, R. M., Steinbusch, H. W. M. and Coulter, J. D. (1981). Serotonergic and peptidergic projections to the spinal cord demonstrated by a combined retrograde HRP histochemical and immunocytochemical staining method. *Brain Res.* **211**, 412–417.

Bowker, R. M., Westlund, K. N. and Coulter, J. D. (1981). Origins of serotonergic projections to the spinal cord in rat: an immunocytochemical-retrograde transport study. *Brain Res.* **226**, 187–200.

Brown, A. G. and Fyffe, R. E. W. (1981). Direct observations on the contacts made between Ia afferent fibres and α-motoneurones in the cat's lumbosacral spinal cord. *J. Physiol.* **313**, 121–140.

Brown, A. G. and Noble, R. (1982). Connexions between hair follicle afferent fibres and spinocervical trace neurones in the cat: the synthesis of receptive fields. *J. Physiol.* **323**, 77–91.

Brown, A. G., House, C. R., Rose, P. K. and Snow, P. J. (1976). The morphology of spinocervical tract neurones in the cat. *J. Physiol.* **260**, 719–738.

Brown, A. G., Rose, P. K. and Snow, P. J. (1980). Dendritic trees and cutaneous receptive fields of adjacent spinocervical tract neurones in the cat. *J. Physiol.* **300**, 429–440.

Brown, A. G., Fyffe, R. E. W., Maxwell, D. J. and Ralston III, H. J. (1982). The morphology of physiologically identified corticospinal axons in the cat. *Soc. Neurosci. Abstracts.* **12**.

Brown, K. T. and Flaming, D. G. (1975). Instrumentation and technique for beveling fine micropipette electrodes. *Brain Res.* **86**, 172–180.

Brown, K. T. and Flaming, D. G. (1977). New microelectrode techniques for intracellular work in small cells. *Neuroscience* **2**, 813–827.

Bunt, Ann H., Lund, R. D. and Lund, Jennifer S. (1974). Retrograde axonal transport of horseradish peroxidase by ganglion cells of the albino rat retina. *Brain Res.* **73**, 215–228.

Bunt, Ann H., Haschke, R. H., Lund, R. D. and Calkins, D. F. (1976). Factors affecting retrograde axonal transport of horseradish peroxidase in the visual system. *Brain Res.* **102**, 152–155.

Burke, R. E., Walmsley, B. and Hodgson, J. A. (1979). HRP anatomy of group Ia afferent contacts on alpha motoneurons. *Brain Res.* **160**, 347–352.

Clark, R. and Ramsey, R. L. (1975). A stereotaxic animal frame with stepping motor-driven micromanipulator. *J. Physiol.* **244**, 5P.

Cobbett, P. and Cottrell, G. A. (1981). Lucifer Yellow injection of hippocampal neurones in the slice preparation. *J. Physiol.* **310**, 9P.

Corson, D. W., Goddman, S. and Fein, A. (1979). An adaptation of the jet stream microelectrode beveler. *Science, N.Y.* **205**, 1302.

Cullheim, S. and Kellerth, J.-O. (1976). Combined light and electron microscopic tracing of neurones, including axons and synaptic terminals, after intracellular injection of horseradish peroxidase. *Neurosci. Lett.* **2**, 307–313.

Cullheim, S. and Kellerth, J.-O. (1978). A morphological study of the axons and recurrent axon collaterals of cat sciatic α-motoneurons after intracellular staining with horseradish peroxidase. *J. Comp. Neurol.* **178**, 537–558.

Deschênes, M. and Landry, P. (1980). Axonal branch diameter and spacing of nodes in the terminal arborization of identified thalamic and cortical neurones. *Brain Res.* **191**, 538–544.

Eide, E. and Källström, T. (1968). Remotely controlled micromanipulator for neurophysiological use. *Acta physiol. scand.* **73**, 2A.

Ekerot, C.-F. and Oscarsson, O. (1981). Prolonged depolarization elicited in Purkinje cell dendrites by climbing fibre impulses in the cat. *J. Physiol.* **318**, 207–221.

Ensor, D. (1978). Another kind of micro-electrode puller. *J. Physiol.* **284**, 28–29P.

Feder, N. and O'Brien, T. P. (1968). Plant microtechnique: some principles and new methods. *Am. J. Bot.* **55**, 123–142.

Futami, T., Shinoda, Y. and Yokata, J. (1979). Spinal axon collaterals of cortico-spinal neurons identified by intracellular injection of horseradish peroxidase. *Brain Res.* **164**, 279–284.

Globus, A., Lux, H. D. and Schubert, P. (1968). Soma dendritic spread of intra-cellularly-injected tritiated glycine in cat spinal motoneurons. *Brain Res.* **11**, 440–445.

Gobel, S., Falls, W. M., Bennett, G. J., Abdelmoumene, M., Hayashi, H. and Humphrey, E. (1980). An EM analysis of the synaptic connections of horseradish peroxidase-filled stalked cells and islet cells in the substantia gelatinosa of adult cat spinal cord. *J. Comp. Neurol.* **194**, 781–807.

Gogan, P., Gueritaud, J. P. and Tyc-Dumont, Suzanne (1983). Comparison of antidromic and orthodromic action potentials of identified motor axons in the cat's brain stem. *J. Physiol.* **335**, 205–220.

Golgi, C. (1873). Sulla struttura della sostanza grigia del cervello. *Gass. Med. Ital. Lombarda* **33**, 244–246.

Gordon, G. and Miller, R. (1969). Identification of cortical cells projecting to the dorsal column nuclei of the cat. *Q. J. Exp. Physiol.* **54**, 85–98.

Graham, R. C. and Karnovsky, M. J. (1966). The early stages of absorption of injected horseradish peroxidase in the proximal tubules of mouse kidney: Ultrastructural cytochemistry by a new technique. *J. Histochem. Cytochem.* **14**, 291–302.

Graybiel, A. M. and Devor, M. (1974). A microelectrophoretic delivery technique for use with horseradish peroxidase. *Brain Res.* **68**, 167–173.

Green, J. D. (1958). A simple microelectrode for recording from central nervous system. *Nature* **182**, 962.

Hanker, J. S., Yates, P. E., Metz, C. B. and Rustioni, A. (1977). A new, specific, sensitive and non-carcinogenic reagent for the demonstration of horseradish peroxidase. *Histochem. J.* **9**, 789–792.

Harper, A. A. and Lawson, S. N. (1982). Physiology and morphology in rat and mouse primary afferent neurones. *J. Physiol.* **327**, 22P.

Hellon, R. F. (1971). The marking of electrode tip positions in nervous tissue. *J. Physiol.* **214**, 12P.

Hess, R. (1932). "Beiträge zur Physiologie des Hirnstammes." Thieme, Leipzig.

Holländer, H. (1970). The section embedding (SE) technique. A new method for the combined light microscopic and electron microscopic examination of central nervous tissue. *Brain Res.* **20**, 39–47.

Hongo, T., Jankowska, E. and Lundberg, A. (1966). Convergence of excitatory and inhibitory action on interneurones in the lumbosacral cord. *Exp. Brain Res.* **1**, 338–358.

Hongo, T., Kudo, N., Yamashita, M., Ishizuka, N. and Mannen, H. (1981). Transneuronal passage of intraaxonally injected horseradish peroxidase (HRP) from group Ib and II fibers into the secondary neurons in the dorsal horn of the cat spinal cord. *Biomed. Res.* **2**, 722–727.

Houchin, R. J., Maxwell, D. J., Fyffe, R. E. W. and Brown, A. G. (1983). Light and electron microscopy of dorsal spinocerebellar tract neurones in the cat: an intra-cellular horseradish study. *Quart. J. exp. Physiol.* **68**, 719–732.

Hultborn, H., Jankowska, E. and Lindström, S. (1971a). Recurrent inhibition from motor axon collaterals of transmission in the Ia inhibitory pathway to moto-neurones. *J. Physiol.* **215**, 591–612.

Hultborn, H., Jankowska, E. and Lindström, S. (1971b). Recurrent inhibition of interneurones monosynaptically activated from Group Ia afferents. *J. Physiol.* **215**, 613–636.

Hultborn, H., Jankowska, E. and Lindström, S. (1971c). Relative contribution from different nerves to recurrent depression of Ia IPSP's in motoneurones. *J. Physiol.* **215**, 637–664.

Jankowska, E. and Lindström, S. (1970). Morphological identification of physiologically defined neurons in the cat spinal cord. *Brain Res.* **20**, 323–326.

Jankowska, E. and Lindström, S. (1971). Morphological identification of Renshaw cells. *Acta physiol. scand.* **81**, 428–430.

Jankowska, E. and Lindström, S. (1972). Morphology of interneurones mediating Ia reciprocal inhibition of motoneurones in the spinal cord of the cat. *J. Physiol.* **226**, 805–823.

Jankowska, E. and Roberts, W. J. (1972a). An electrophysiological demonstration of the axonal projections of single spinal interneurones in the cat. *J. Physiol.* **222**, 597–622.

Jankowska, E. and Roberts, W. J. (1972b). Synaptic actions of single interneurones mediating reciprocal Ia inhibition of motoneurones. *J. Physiol.* **222**, 623–642.

Jankowska, E., Rastad, J. and Westman, J. (1976). Intracellular application of horseradish peroxidase and its light and electron microscopical appearance in spinocervical tract cells. *Brain Res.* **105**, 557–562.

Jochem, W. J., Light, A. R. and Smith, D. (1981). A high voltage electrometer for recording and iontophoresis with fine-tipped, high resistance microelectrodes. *J. Neurosci. Meth.* **3**, 261–269.

Kater, S. B. and Nicholson, C. (1973). "Intracellular Staining Techniques in Neurobiology". Springer-Verlag, Berlin.

Kellerth, J. O. (1973). Intracellular staining of cat spinal motoneurons with Procion Yellow for ultrastructural studies. *Brain Res.* **50**, 415–418.

Kelly, J. P. and Van Essen, D. C. (1974). Cell structure and function in the visual cortex of the cat. *J. Physiol.* **238**, 515–547.

Kerkut, G. A. and Walker, R. J. (1962). Marking individual nerve cells through electrophoresis of ferrocyanide from a microelectrode. *Stain Technol.* **37**, 217–219.

Kitai, S. T., Kocsis, J. D., Preston, R. J. and Sugimori, M. (1976). Monosynaptic inputs to caudate neurons identified by intracellular injection of horseradish peroxidase. *Brain Res.* **109**, 601–606.

Knowles, W. D., Funch, P. G. and Schwartzkroin, P. A. (1982). Electrotonic and dye coupling in hippocampal CA1 pyramidal cells *in vitro. Neuroscience* **7**, 1713–1722.

Kravitz, E. A., Stretton, A. O. W., Alvarez, J. and Furshpan, E. J. (1968). Determination of neuronal geometry using an intracellular dye injection technique. *Fedn Proc. Fedn Am. Socs exp. Biol.* **27**, 749.

Kreutzberg, G. W., Schubert, P. and Lux, H. D. (1975). Neuroplasmic transport in axons and dendrites. *In* "Golgi Centennial Symposium: Perspectives in Neurobiology" (M. Santini, ed.), pp. 161–166. Raven Press, New York.

Laporte, Y., Lundberg, A. and Oscarsson, O. (1956). Functional organization of the dorsal spino-cerebellar tract in the cat. II. Single fibre recording in Fleschig's fasciculus on electrical stimulation of various peripheral nerves. *Acta physiol. scand.* **36**, 188–203.

LaVail, J. H. and LaVail, M. M. (1972). Retrograde axonal transport in the central nervous system. *Science* **176**, 1416–1417.

LaVail, J. H. and LaVail, M. M. (1974). The retrograde intra-axonal transport of horseradish peroxidase in the chick visual system: a light and electron microscopic study. *J. Comp. Neurol.* **157**, 303–358.

Lederer, W. J., Spindler, A. J. and Eisner, D. A. (1979). Thick slurry bevelling. A new technique for bevelling extremely fine microelectrodes and micropipettes. *Pflügers Arch.* **381**, 287–288.

Light, A. R.(1983). Spinal projections of physiologically identified axons from the rostral medulla of cats. "Advances in Pain Research and Therapy" (J. R. Bonica, U. Lindblom and A. Iggo, eds), Vol. 5, pp. 273–280. Raven Press, New York.

Light, A. R. and Durkovic, R. G. (1976). Horseradish peroxidase: An improvement in intracellular staining of single, electrophysiologically characterized neurons. *Exp. Neurol.* **53**, 847–853.

Light, A. R. and Perl, E. R. (1979). Spinal termination of functionally identified primary afferent neurons with slowly conducting myelinated fibers. *J. Comp. Neurol.* **186**, 133–150.

Light, A. R., Trevino, D. A. and Perl, E. R. (1979). Morphological features of functionally defined neurons in the marginal zone and substantia gelatinosa of the spinal dorsal horn. *J. Comp. Neurol.* **186**, 151–172.

Light, A. R., Réthelyi, M. and Perl, E. R. (1981). Ultrastructure of functionally identified neurones in the marginal zone and the substantia gelatinosa. *In* "Spinal Cord Sensation" (A. G. Brown and M. Réthelyi, eds), pp. 97–102. Scottish Academic Press, Edinburgh.

Ling, G. and Gerard, R. W. (1949). The normal membrane potential of frog sartorius muscle fibers. *J. Cell. comp. Physiol.* **34**, 383–396.

Llinas, R. and Nicholson, C. (1971). Electrophysiological properties of dendrites and somata in alligator Purkinje cells. *J. Neurophysiol.* **34**, 532–551.

Lux, H. D., Schubert, P. and Kreutzberg, G. W. (1970a). Direct matching of morphological and electrophysiological data in cat spinal motoneurons. *In* "Excitatory Synaptic Mechanisms (P. Anderson and J. K. S. Jansen, eds), pp. 189–198. Universitetsforlaget, Oslo.

Lux, H. D., Schubert, P., Kreutzberg, G. W. and Globus, A. (1970b). Excitation and axonal flow. Autoradiographic study on motoneurons intracellularly injected with ^{3}H-amino acid. *Exp. Brain Res.* **10**, 197–204.

Martin, K. A. C. and Whitteridge, D. (1982). The morphology, function and intracortical projections of neurones in area 17 of the cat which receive monosynaptic input from the lateral geniculate nucleus. *J. Physiol.* **328**, 37P.

Maxwell, D. J., Bannatyne, B. A., Fyffe, R. E. W. and Brown, A. G. (1982). Ultrastructure of hair follicle afferent fibre terminations in the spinal cord of the cat. *J. Neurocytol.* **11**, 571–582.

Maxwell, D. J., Fyffe, R. E. W. and Brown, A. G. (1982). Fine structure of spinocervical tract neurones and the synaptic boutons in contact with them. *Brain Res.* **233**, 394–399.

Mense, S., Light, A. R. and Perl, E. (1981). Spinal terminations of subcutaneous high-threshold mechanoreceptors. *In* "Spinal Cord Sensation" (A. G. Brown and M. Réthelyi, eds), pp. 79–86. Scottish Academic Press, Edinburgh.

Mesulam, M.-M. (1978). Tetramethyl benzidine for horseradish peroxidase neurohistochemistry. A non-carcinogenic blue reaction-product with superior sensitivity for visualizing neural afferents and efferents. *J. Histochem. Cytochem.* **26**, 106–117.

Metz, C. B., Kavookjian, A. M. and Light, A. R. (1982). Techniques for HRP intracellular staining of neural elements for light and electron microscopic analyses. *J. Electrophysiol. Tech.* **9**, 151–163.

Molony, V., Steedman, Wilma M., Cervero, F. and Iggo, A. (1981). Intracellular marking of identified neurones in the superficial dorsal horn of the cat spinal cord. *Q. J. Exp. Physiol.* **66**, 211–223.

Ogden, T. E., Citron, M. C. and Pierantoni, R. (1978). The jet stream microbeveler: an inexpensive way to bevel glass micropipettes. *Science, N.Y.* **201**, 469–470.

Peters, A., Palay, S. L. and Webster, H. deF. (1976). "The Fine Structure of the Nervous Systems: the Neurons and Supporting Cells". Saunders, Philadelphia.

Phillips, C. G., and Porter, R. (1977). "Corticospinal Neurones: Their Role in Movement" pp. 54–57. Academic Press, London, Orlando and New York.

Pitman, R. M. C. (1984). "Intracellular Staining in Invertebrates". Academic Press, London, Orlando and New York.

Pitman, R. M. C., Tweedle, C. D. and Cohen, M. J. (1972a). Branching of central neurons: intracellular cobalt injection for light and electron microscopy' *Science* **176**, 412–414.

Pitman, R. M. C., Tweedle, C. D. and Cohen, M. J. (1972b). Electrical responses of insect central neurons: augmentation by nerve section or colchicine. *Science* **178**, 507–509.

Purves, R. D. (1981). "Microelectrode Methods for Intracellular Recording and Ionophoresis". Academic Press, London, Orlando and New York.

Purves, R. D. and McMahan, U. J. (1972). The distribution of synapses on a physiologically identified motor neuron in the central nervous system of the leech: an electron microscopic study after the injection of the fluorescent dye Procion Yellow. *J. Cell Biol.* **55**, 205–220.

Rastad, J., Jankowska, E. and Westman, J. (1977). Arborization of intial axon collaterals of spinocervical tract cells stained intracellularly with horseradish peroxidase. *Brain Res.* **135**, 1–10.

Redman, S., Walmsley, B. (1983). The time course of synaptic potentials evoked in cat spinal motoneurons at identified Group Ia synapses. *J. Physiol.* **343**, 117–133.

Remler, M. P., Selverston, A. I. and Kennedy, D. (1968). Lateral giant fibers of crayfish: Location of somata by dye injection. *Science* **162**, 281–283.

Réthelyi, M., Light, A. R. and Perl, E. R. (1982). Synaptic complexes formed by functionally defined primary afferent units with fine myelinated fibers. *J. Comp. Neurol.* **207**, 381–393.

Reynolds, E. D. (1963). The use of lead citrate at high pH as electron-opaque stain in electron microscopy. *J. Cell Biol.* **17**, 208–212.

Rose, P. K. (1981). Distribution of dendrites from biventer cervicis and complexus motoneurons stained intracellularly with horseradish peroxidase in the adult cat. *J. Comp. Neurol.* **197**, 395–409.

Rose, P. K. and Richmond, F. J. R. (1981). White-matter dendrites in the upper cervical spinal cord of the adult cat: a light and electron microscopic study. *J. Comp. Neurol.* **199**, 191–203.

Schubert, P. (1974). Transport in Dendriten einzelner Motoneurone. *Bull. Schweiz. Akad. Med. Wiss.* **30**, 56–65.

Somogyi, P. and Smith, A. D. (1979). Projection of neostriatal spiny neurons to the substantia nigra. Application of a combined Golgi-staining and horseradish peroxidase transport procedure at both light and electron microscopic levels. *Brain Res.* **178**, 3–15.

Somogyi, P., Hodgson, A. J. and Smith, A. D. (1979). An approach to tracing neuron networks in the cerebral cortex and basal ganglia. Combination of Golgi staining, retrograde transport of horseradish peroxidase and anterograde degeneration of synaptic boutons in the same material. *Neuroscience* **4**, 1805–1852.

Snow, P. J., Rose, P.K. and Brown, A. G. (1976). Tracing axons and axon collaterals of spinal neurons using intracellular injection of horseradish peroxidase. *Science, N.Y.* **191**, 312–313.

Stewart, W. W. (1978). Functional connections between cells as revealed by dye-coupling with a highly fluorescent naphthalimide tracer. *Cell* **14**, 741–759.

Stretton, A. O. W. and Kravitz, E. A. (1968). Neuronal geometry: determination with a technique of intracellular dye injection. *Science* **162**, 132–134.

Stretton, A. O. W. and Kravitz, E. A. (1973). Intracellular dye injection: the selection of Procion Yellow and its application in preliminary studies of neuronal geometry in the lobster nervous system. *In* "Intracellular Staining in Neurobiology" (S. B. Kater and C. Nicholson, eds), pp. 21–40. Springer-Verlag, Berlin.

Thomas, R. C. (1978). *Ion-sensitive Intracellular Microelectrodes.* Academic Press, London, Orlando and New York.

Thomas, R. C. and Wilson, U. J. (1966). Marking single neurons by staining with intracellular recording electrodes. *Science* **151**, 1538–1539.

Tracey, D. J. and Walmsley, B. (1982). Anatomy of the connection between identified primary afferents and neurones of the dorsal spinocerebellar tract. *Proc. Aust. Physiol. Pharmac Soc.* **13**(2), 132P.

Van Essen, D. and Kelly, J. (1973). Morphological identification of simple, complex and hypercomplex cells in the visual cortex of the cat. *In* "Intracellular Staining in Neurobiology" (S.B. Kater and C. Nicholson, eds), pp. 189–198. Springer-Verlag, Berlin.

Van Orden, L. S. (1973). Principles of fluorescence microscopy and photomicrography with applications to Procion Yellow. *In* "Intracellular Staining in Neurobiology" (S. B. Kater and C. Nicholson, eds), pp. 61–69. Springer-Verlag, Berlin.

Warr, W. B., de Olmos, J. S. and Heimer, L. (1981). Horseradish peroxidase: The basic procedure. *In* "Neuroanatomical Tract-Tracing Methods (L. Heimer and M. J. RoBards, eds), Plenum Press, New York.

Winsbury, G. (1956). Machine for the production of microelectrodes. *Rev. scient. Instrum.* **27**, 514–516.

Yezierski, R. P., Bowker, R. M., Kevetter, G. A., Westlund, K. N., Coulter, J. D. and Willis, W. D. (1982). Serotonergic projections to the caudal brain stem: a double label study using horseradish peroxidase and serotonin immunocytochemistry. *Brain Res.* **239**, 258–264.

Zieglgänsberger, W. and Reiter, C. (1974). Interneuronal movement of Procion Yellow in cat spinal neurons. *Exp. Brain Res.* **20**, 527–530.

Index

Page numbers in *italics* refer to figures, those in **bold** type to tables